STRONG GRAVITATIONAL LENSING IN THE ERA OF BIG DATA

IAU SYMPOSIUM 381

COVER ILLUSTRATION:

Webb's First Deep Field (NIRCam Image) (Credit: NASA, ESA, CSA, STScI)

IAU SYMPOSIUM PROCEEDINGS SERIES

INTERNATIONAL ASTRONOMICAL UNION

UNION ASTRONOMIQUE INTERNATIONALE

International Astronomical Union

IAU

STRONG GRAVITATIONAL LENSING IN THE ERA OF BIG DATA

PROCEEDINGS OF THE 381st SYMPOSIUM OF THE INTERNATIONAL ASTRONOMICAL UNION OTRANTO, ITALY 19–23 JUNE 2023

Edited by

HANNAH STACEY

European Southern Observatory;
Max Planck Institute for Astrophysics, Germany

ALESSANDRO SONNENFELD

Shanghai Jiao Tong University

and

CLAUDIO GRILLO

University of Milan; INAF-IASF Milan, Italy

CAMBRIDGE
UNIVERSITY PRESS

CAMBRIDGE UNIVERSITY PRESS
University Printing House, Cambridge CB2 8BS, United Kingdom
1 Liberty Plaza, Floor 20, New York, NY 10006, USA
10 Stamford Road, Oakleigh, Melbourne 3166, Australia

First published 2024

Printed in Great Britain by Henry Ling Limited, The Dorset Press, Dorchester, DT1 1HD

Typeset in System LaTeX 2ε

A catalogue record for this book is available from the British Library of Congress Cataloguing in Publication data

ISBN 9781009398992 hardback
ISSN 1743-9213

Table of Contents

Chapter 5

Preface

Strong gravitational lensing is a powerful and mature technique for probing galaxies and the Universe as a whole. In the past twenty years, strong lensing observations have enabled unique studies, such as measurements of the dark matter distribution in galaxies and galaxy clusters, detections of substructure in galaxies, measurements of the expansion rate of the Universe with time-delay lenses, and high-resolution analyses of strongly lensed high-redshift galaxies. While these efforts have improved our understanding of galaxy evolution and cosmology, there are still open questions that strong lensing observations can help to address. What is the nature of dark matter and dark energy? Is the history of the Universe well described by a flat Λ cold dark matter model? The answers to these questions have profound fundamental physics implications.

The field of strong lensing is about to be revolutionised by the advent of new observational facilities, such as Euclid, the Rubin Observatory, the Roman Space Telescope and the Chinese Space Station Telescope. These telescopes and their associated surveys are expected to lead to the discovery of around 100000 new strong lenses, an increase of more than two orders of magnitude with respect to the current sample size of confirmed lenses. Such a large amount of data gives us the potential for carrying out strong lensing studies with very high precision. At the same time, however, it poses new challenges: to fully take advantage of this improvement in precision, it is necessary for the accuracy of the models used to interpret these data to improve as well. Furthermore, traditional analysis methods, based on the detailed study of a few select systems, do not scale well to very large sample sizes. With this symposium, we brought together observational and theoretical researchers in the community to review the progress of the field and develop a roadmap for the new era of strong gravitational lensing.

The goals of the symposium were

- to understand the preparedness of the community to analyse and adapt to forthcoming big data;
- to assess the progress of the field in regard to precision lens modelling and modelling of high-angular-resolution data, and to understand what developments are needed;
- to find synergies between observations and simulations of dark matter and the high-redshift Universe;
- to determine the scientific questions that strong lensing will address over the next decade and foster collaboration.

Four major subjects were covered during the week: cosmology, dark matter, galaxies and galaxy clusters, and high-redshift sources. Each subject had dedicated sessions with invited and contributed talks, as well as posters and discussion sessions. Sessions dedicated to the same subject were spread out over multiple days to encourage synergies between the different science aspects and methodology.

The proposed symposium date of June, 2023, was auspicious as it preceded the first data release from flagship survey facilities. In the past few years, many new astronomers have entered the field, started groups and introduced fresh and innovative ideas. During the symposium, these researchers presented the state-of-the-art techniques they developed to resolve structure in the early Universe, test the nature of dark matter, test fundamental physics and efficiently analyse forthcoming big data sets, demonstrating that the strong lensing community is well-placed to address critical scientific questions in astrophysics. A selection of the work presented is summarised in these proceedings.

Hannah Stacey
Alessandro Sonnenfeld
Claudio Grillo

Editors

Hannah Stacey (European Southern Observatory; Max Planck Institute for Astrophysics, Germany)
Alessandro Sonnenfeld (Shanghai Jiao Tong University)
Claudio Grillo (University of Milan; INAF-IASF Milan, Italy)

Organizing Committee

Scientific Organizing Committee

Local Organizing Committee

List of Participants

Aristeidis Amvrosiadis	Durham University
Dominique Sluse	University of Liège
Uros Mestric	University of Milano
Ashish Kumar Meena	Ben-Gurion University of the Negev
Irham Taufik Andika	Technical University of Munich
Patrick Kamieneski	Arizona State University
Cristiana Spingola	INAF Institute for Radioastronomy
Q.Daniel Wang	University of Massachusetts Amherst
Raquel Forés-Toribio	University of Valencia
Carina Fian	University of Valencia
Edoardo Borsato	University of Padua
Favio Neira	École Polytechnique Fédérale de Lausanne
Felipe Ávila	Universidad Valparaíso
Aprajita Verma	Oxford University
John McKean	University of Groningen; Netherlands Institute for Radio Astronomy
Andrea Bolamperti	University of Padua; European Southern Observatory
Grasiele Romanzini Bezerra	Brazilian Center for Research in Physics
Hannah Turner	Durham University
Martin Millon	Stanford University
Chin Yi Tan	University of Chicago
Tania Barone	Swinburne University of Technology
Nan Li	National Astronomical Observatories, Chinese Academy of Sciences
Devon Williams	University of California Los Angeles
Javier Alejandro Acevedo Barroso	École polytechnique fédérale de Lausanne
Karl Glazebrook	Swinburne University of Technology
Nandini Sahu	University of New South Wales
Jimena Gonzalez	University of Wisconsin-Madison
Kamal Bora	Swinburne University of Technology; Indian Institute of Technology
Karina Rojas	University of Portsmouth
Alejandra Melo	Max Planck Institute for Astrophysics
Bharath Chowdhary Nagam	University of Groningen
João Paulo França	Brazilian Center for Research in Physics
Hareesh Thuruthipilly	National Center for Nuclear Research
Philip Holloway	Oxford University
Timo Anguita	Universidad Andrés Bello
Daniel Ballard	University of Portsmouth
Dorota Bayer	Swinburne University of Technology
Devon Powell	Max Planck Institute for Astrophysics
Conor O'Riordan	Max Planck Institute for Astrophysics
Di Wen	University of Groningen
Joshua Fagin	City University of New York
Birendra Dhanasingham	University of New Mexico Albuquerque
Chris Fassnacht	University of California Davis
Daniel Gilman	University of Toronto
Georgios Vernardos	City University of New York
Sergei Gleyzer	University of Alabama
Sebastian Wagner-Carena	Stanford University
Tyler Hughes	Swinburne University of Technology

Wolfgang Enzi	University of Portsmouth
Dan Ryczanowski	University of Birmingham
Hakon Dahle	University of Oslo
Shawn Knabel	University of California Los Angeles
Justin Pierel	Space Telescope Science Institute
Sydney Erickson	Stanford University
Lyne Van de Vyvere	University of Liège
Matthew Gomer	University of Liège
Nikki Arendse	Oskar Klein Centre, Stockholm University
Raoul Canameras	Max Planck Institute for Astrophysics
Graham Smith	University of Birmingham
Martin Makler	Universidad Nacional de San Martín
Satadru Bag	Max Planck Institute for Astrophysics
Carlos Melo-Carneiro	Universidade Federal do Rio Grande do Sul
Giacomo Queirolo	Max Planck Institute for Extraterrestrial Physics; Origins Excellence Cluster; Ludwig Maximilian University of Munich
Henry Best	City University of New York
James Chan	City University of New York
Shuaibo Geng	National Center for Nuclear Research
Giuseppe Angora	University of Ferrara
Giovanni Granata	University of Milan
Stefan Schuldt	University of Milan
Han Wang	Max Planck Institute for Astrophysics
Pietro Bergamini	University of Milan
Davide Abriola	University of Milan
Lukas Furtak	Ben-Gurion University of the Negev
Raven Gassis	University of Cincinnati
Lorenzo Bazzanini	University of Ferrara
Sangjun Cha	Yonsei University
Jackson ODonnell	University of California Santa Cruz
Cameron Lemon	École Polytechnique Fédérale de Lausanne
Alessandro Sonnenfeld	Shanghai Jiao Tong University
Ana Acebron	University of Milan
Ana Sainz de Murieta	University of Portsmouth
Anna Nierenberg	University of California Merced
Anowar Shajib	University of Chicago
Bridget Ierace	City University of New York
Claudio Grillo	University of Milan
Francesca Rizzo	Cosmic Dawn Center
Gabriel Bartosch Caminha	Technical University of Munich; Max Planck Institute for Astrophysics
Giulia Despali	University of Bologna
Hannah Stacey	Max Planck Institute for Astrophysics
James Nightingale	Durham University
Kenneth Wong	National Astronomical Observatory of Japan
Luke Weisenbach	University of Portsmouth
Masamune Oguri	Chiba University
Massimo Meneghetti	INAF-OAS Bologna
Matt O'Dowd	City University of New York
Piero Rosati	University of Ferrara
Quinn Minor	City University of New York
Richard Massey	Durham University

Rui Li	Chinese Academy of Sciences
Russell Smith	Durham University
Samuel Lange	Durham University
Sherry Suyu	Technical University of Munich; Max Planck Institute for Astrophysics
Simon Birrer	Stony Brook University
Thomas Collett	University of Portsmouth
Tian Li	University of Portsmouth
Tommaso Treu	University of California Los Angeles
Wenshuo Xu	Tsinghua University
Yan Liang	Tsinghua University
Brian Welch	NASA Goddard Space Flight Center
Shude Mao	Tsinghua University China
Zizhao He	Chinese Academy of Sciences
Luca Di Mascolo	Università degli Studi di Trieste
Fabio Convenga	Karlsruhe Institute of Technology

Chapter 1

Strong Gravitational Lensing in the Era of Big Data
Proceedings IAU Symposium No. 381, 2024
H. Stacey, A. Sonnenfeld & C. Grillo, eds.
doi:10.1017/S1743921323003903

Strong Lensing by Galaxies: Past Highlights, Current Status, and Future Prospects

Anowar J. Shajib

Department of Astronomy & Astrophysics, The University of Chicago, Chicago, IL 60637, USA
Kavli Institute for Cosmological Physics, University of Chicago, Chicago, IL 60637, USA
NHFP Einstein Fellow, Email: ajshajib@uchicago.edu

Abstract. Galaxy-scale strong lensing is a powerful tool in Astrophysics and Cosmology, enabling studies of massive galaxies' internal structure, their formation and evolution, stellar initial mass function, and cosmological parameters. In this conference proceeding, we highlight key findings from the past decade in astrophysical applications of strong lensing at the galaxy scale. We then briefly summarize the present status of discovery and analyses of new samples from recent or ongoing surveys. Finally, we offer insights into anticipated developments in the upcoming era of big data shaping the future of this field, thanks to the Rubin, *Euclid*, and *Roman* observatories.

Keywords. Strong lensing, galaxies

1. Introduction

Strong lensing by galaxies provides a valuable probe of the internal structure of the lensing galaxies (Shajib et al. 2022). Since elliptical galaxies are the most common type of lensing galaxies, studies of strong lensing galaxies have mainly focused on massive ellipticals. The internal structure of elliptical galaxies is shaped through cosmic time by the initial baryonic infall, star formation, subsequent outflow induced by baryonic feedback mechanisms, and hierarchical formation through mergers. All of these baryonic processes are also accompanied by adiabatic contraction and expansion of the dark matter. Thus, constraining the internal structure of galaxies, that is, the distribution of baryonic and dark matter, at different redshifts can shed light on their formation and evolutionary history. Stellar dynamics, especially the spatially resolved kind, of local elliptical galaxies have provided great insight into their formation and evolution (Cappellari 2016). However, strong lensing offers unique complementarities to stellar dynamics in providing more accessible and informative data for galaxies at redshift $z \gtrsim 0.5$ and by providing an independent tracer of the mass to break degeneracies inherent to both probes, when combined.

Since the first discovery of strong lensing systems at the galaxy scale in the eighties, the field has come a long way in efficiently harnessing the rich information in large samples of strong lenses. Thanks to these large samples, for example, the Sloan Lens ACS Survey (SLACS; Bolton et al. 2006) and the Cosmic Lens All Sky Survey (CLASS; Myers et al. 1995), the internal structure of elliptical galaxies at $z \sim 0.3$ were unveiled with significant precision in the 2000s. However, at the beginning of the past decade (2010s), Treu (2010) posed the following two main outstanding questions to be solved by strong lensing studies of galaxies:

(1) How do luminous and dark matter density profiles evolve over cosmic time?
(2) Does the dark matter density profile universally follow the Navarro–Frenk–White (NFW; Navarro et al. 1996, 1997) profile predicted by cosmological simulations?

In this proceeding, we briefly introduce the progress made on these two questions over the past decade (Section 2). We then introduce the current status of the discovery of larger lens samples and their analyses (Section 3), before concluding the proceeding with a discussion on the prospects for the forthcoming era of big data (Section 4). The readers are invited to see Shajib et al. (2022) for a more comprehensive review of the topic.

2. Highlights of past results

In this section, we present some major results from the literature on the internal structure of elliptical galaxies (Section 2.1) and the stellar initial mass function (IMF; Section 2.2).

2.1. *Internal structure of elliptical galaxies*

Numerous strong lensing studies found that the total density profile in elliptical galaxies is well approximated by a power-law profile, that is, $\rho(r) \propto r^{-\gamma}$. These studies also showed that the power-law profile is nearly isothermal, that is, $\gamma \sim 2$. This result holds for the sample-mean of the logarithmic slope γ constrained by both joint lensing–dynamics analyses [e.g., 2.08 ± 0.03 from the SLACS (Auger et al. 2009), 2.11 ± 0.02 from the BOSS Emission-Line Lens Survey (BELLS; Bolton et al. 2012), 2.05 ± 0.06 from the Strong Lensing Legacy Survey (SL2S; Sonnenfeld et al. 2013)] and strong-lensing-only analyses [e.g., 2.08 ± 0.03 from SLACS (Shajib et al. 2020), and 2.09 ± 0.03 from SLACS and BELLS (Etherington et al. 2022)]. These analyses also found intrinsic scatters in γ between $0.13 - 0.19$.

The logarithmic slope γ constrained by the aforementioned studies can be compared to those from cosmological hydrodynamical simulations with various prescriptions of baryonic physics to learn about the particular baryonic physics that has been at play in shaping the elliptical galaxies. For example, Mukherjee et al. (2021) rule out scenarios such as no AGN feedback and, lower-viscosity AGN accretion, or environment-dependent stellar feedback (Fig. 1).

The evolution of the slope of the mass density profile over redshift can additionally provide further insights into the baryonic feedback processes and merger types that played crucial roles during the lifespan of elliptical galaxies at $z < 1$. Hydrodynamic simulations, such as IllustrisTNG and Magneticum, show that gas-poor mergers leave the logarithmic slope in elliptical galaxies unchanged, or make it slightly shallower with decreasing redshift (Fig. 2; Remus et al. 2017; Wang et al. 2020). Conversely, gas-rich mergers would have made the average logarithmic slope steeper. In apparent contrast with the simulations, joint lensing–dynamics studies found that the logarithmic slope γ gets steeper with decreasing redshift, thus favoring the gas-rich merger scenario (Ruff et al. 2011; Bolton et al. 2012). However, this discrepancy between simulations and observations can also be explained by systematic effects, such as selection effect (Sonnenfeld et al. 2015), projection effect (Remus et al. 2017), and parametrization of the stellar anisotropy (Xu et al. 2017).

Furthermore, Etherington et al. (2022) found no correlation in the logarithmic slopes obtained from lensing-only analysis and joint lensing–dynamic analysis. These authors, therefore, argue that the true mass distribution in elliptical galaxies must deviate from a pure power-law form. To allow more radial flexibility than a power law, several studies modeled the mass distribution in elliptical galaxies using a two-component model

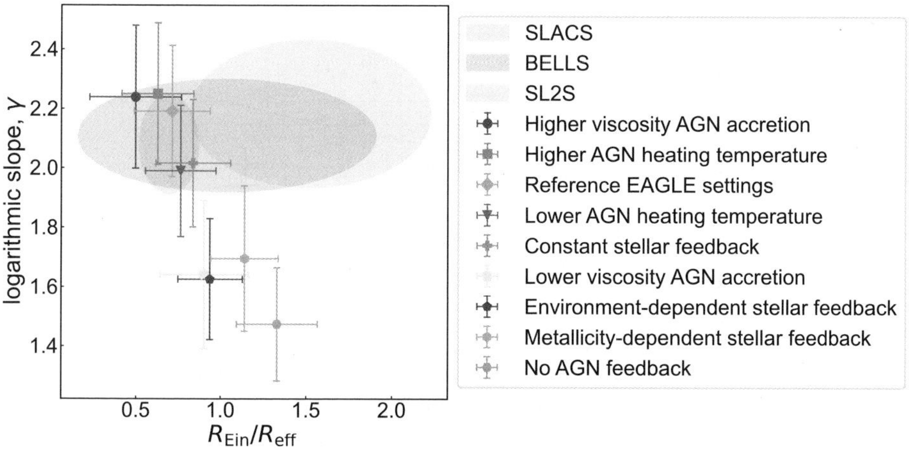

Figure 1. Comparison of the constrained logarithmic slope γ and Einstein radius $R_{\rm Ein}$ between observations (shaded ellipses) and predictions from the EAGLE simulation with various baryonic physics prescriptions (points with errorbars). Some prescriptions, such as the no-AGN feedback model (grey point), can be ruled out. Figure re-illustrated from Mukherjee et al. (2021).

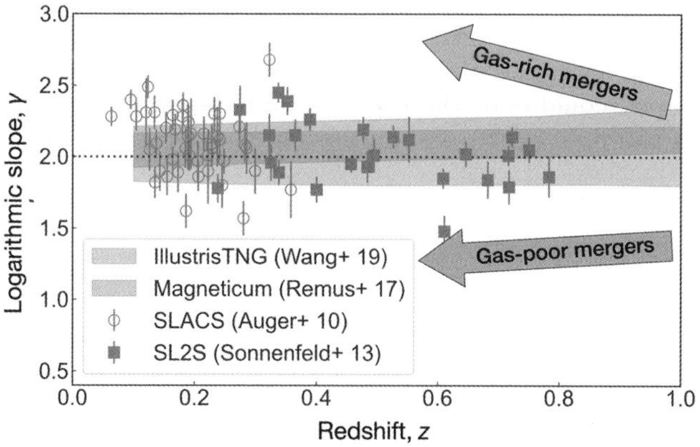

Figure 2. Evolution of the average logarithmic slope γ with redshift z. The green and pink arrows indicate the direction of evolution in γ for gas-rich and gas-poor mergers, respectively. Simulations (orange and blue shaded stripes) indicate gas-poor mergers make the slope slightly shallower with decreasing redshift. However, joint lensing–dynamics observations indicate the opposite trend favoring the gas-rich merger scenario.

with one mass profile describing the dark matter halo (typically with an NFW profile) and the other describing the stellar mass distribution. However, some disagreement exists in the literature about the departure from the "vanilla" (i.e., not contracted nor expanded) NFW profile and the presence of a mass-to-light ratio gradient in the stellar mass distribution. Dutton and Treu (2014) found the vanilla NFW profile fits the lensing and dynamics data, whereas scenarios with adiabatic contraction or expansion do not. However, Oldham and Auger (2018) found a bimodality in their sample with the inner slope shallower (expanded) than the vanilla NFW in one mode and steeper (contracted) in the other. In contrast, Sonnenfeld et al. (2018) found that a mass-to-light ratio gradient in the stellar mass profile, not a departure from the vanilla NFW profile, fits better the combined dataset from strong lensing, dynamics, and weak lensing. Finally, Shajib et al. (2021) ruled our contraction or expansion in the dark matter halos and found

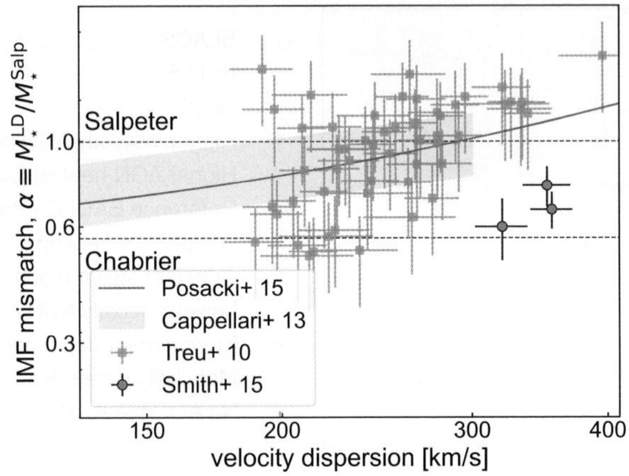

Figure 3. Compilation of results on the stellar IMF from lensing and dynamics. The IMF mismatch parameter $\alpha = 1$ represents the Salpeter IMF. The majority of constraints from lensing and dynamics indicate a trend in the IMF with the velocity dispersion (purple line, Posacki et al. 2015). However, some measurements, for example, the blue points from Smith et al. (2015), disagree with this trend. Figure re-illustrated from Smith et al. (2015).

no strong evidence in favor of a mass-to-light ratio gradient in the stellar mass from the combination of strong lensing, dynamic, and weak lensing data. However, these discrepancies in the literature could potentially arise from different treatment stellar anisotropy parameterizations and light profile assumptions in the dynamical modeling.

2.2. *Stellar IMF*

Most strong lensing studies found the stellar IMF to be heavier (i.e., Salpeter IMF; e.g., Treu et al. 2010; Spiniello et al. 2011; Sonnenfeld et al. 2012) than that found in the Milky Way (i.e., Chabrier IMF). These results also agree very well with several non-lensing constraints on the IMF, such as those based on stellar dynamics and the stellar population synthesis method (Cappellari et al. 2012; La Barbera et al. 2013; Spiniello et al. 2014). Fig. 3 illustrates this agreement between lensing-based and dynamics-based studies and the apparent dependency of the IMF on the velocity dispersion (Posacki et al. 2015).

However, there are also reports of a lighter IMF, such as the Chabrier IMF, from lensing-based studies (e.g., Ferreras et al. 2010; Smith et al. 2015; Sonnenfeld et al. 2019). However, systematics stemming from parameterizations of the dark matter profile, the mass-to-light-ratio gradient, or the stellar anisotropy cannot be ruled out as the source of this disagreement.

Despite much progress with multiple large samples of lenses over the last decade, the two questions posed in Section 1 are yet to be answered definitively. To achieve that goal in the future, larger statistical samples, combining strong lensing, weak lensing, and dynamics, and adopting models with more radial flexibility than the power law would be essential. In the next section, we describe the current efforts in these aspects.

3. Presently emerging samples and analyses

One way to achieve a larger statistical sample than the previously available ones is to create a "super-sample" assembled from archival samples. Following this way, Project Dinos (PI: Shajib) was created to study the evolution of elliptical galaxies' internal structure at $z < 1$ using as many suitable archival samples as possible. From this project, Tan et al. (in preparation) will present an analysis of the "super-sample"

(SLACS+SL2S+BELLS) with available archival *Hubble Space Telescope* (*HST*) data and ground-based velocity dispersion measurements.

Recently, lots of new samples have also been discovered, primarily using machine-learning techniques applied on several ground-based surveys (e.g., Agnello et al. 2018; Delchambre et al. 2019; Wong et al. 2022; Lemon et al. 2023). An emerging sample with high-resolution *HST* imaging and high-quality ground-based velocity dispersion measurements is the Astro3D Galaxy Evolution with Lenses (AGEL) sample (Tran et al. 2022). This sample primarily consists of newly discovered strong lenses from the Dark Energy Survey data (Jacobs et al. 2019b,a), and includes mostly galaxy-scale lenses with a subset of group-scale lenses.

Recently, two *HST* Schedule Gap Program has been approved to take imaging data of the newly discovered lens systems from different surveys: one for galaxy–galaxy lenses (HST-SNAP-17307, PIs: Tran & Shajib), and the other for lensed quasar systems (HST-SNAP-17308, PI: Lemon). These two programs, combined, will obtain high-resolution *HST* imaging of ~450–600 galaxy-scale lenses over the next 3–5 years.

4. Future prospects in the era of big data

With the imminent launch of the Vera C. Rubin Observatory's Legacy Survey of Space and Time (LSST), and from current and future space-based surveys such as the *Euclid* and the *Roman Space Telescope*, even larger samples with size $\mathcal{O}(10^4) - \mathcal{O}(10^5)$ will be discovered (Oguri and Marshall 2010; Collett 2015). Discovering these lenses would primarily utilize machine learning techniques with more novel algorithms currently being developed (e.g., Akhazhanov et al. 2022). However, analyzing such unprecedentedly large samples will also pose a computational challenge. One way to tackle this challenge is to use machine learning for lensing parameter extraction from the data, which is currently an active area of research (e.g., Hezaveh et al. 2017; Morningstar et al. 2019; Schuldt et al. 2021; Adam et al. 2022; Poh et al. 2022; Mishra-Sharma and Yang 2022; Biggio et al. 2023). However, depending on specific scientific requirements, the conventional forward modeling approach would still be favorable for a subset of all the new lenses. This subset of lenses will still be too large (e.g., $\mathcal{O}(10^3)$) for feasible analysis with the traditional method that requires a considerable amount of fine-tuning by a human modeler on a case-by-case basis. Several efforts are underway to develop automated lens modeling pipelines, e.g., using LENSTRONOMY (Birrer and Amara 2018; Birrer et al. 2021; Shajib et al. 2019, 2021; Schmidt et al. 2022), PYAUTOLENS (Nightingale et al. 2021), using GLEE (Ertl et al. 2023).

As the statistical precision grows tighter with these future large samples, systematic effects such as the selection function will become increasingly important, if not already. Using simulation, Sonnenfeld et al. (2023) show that strong lensing samples can be biased in the IMF mismatch parameter by ~10% and in the dark matter inner slope by ~5% (Fig. 4). However, the selection effect from more factors, such as the halo concentration, environment, viewing angles, etc., could still be investigated by future studies. Accurately estimating the selection bias from these factors would allow future analyses of strong lens samples to correct the potential selection biases.

Although the last decade saw a large amount of progress in discovering new lens samples and in novel modeling and analysis techniques, several key questions about the internal structure and evolution of massive elliptical galaxies still lack definitive answers. This decade is expected to revolutionize all applications of strong lensing at the galaxy scale due to the two or more orders of magnitude increase in the lens sample size. Thus, Rubin, *Euclid*, and *Roman* observatories will play key roles in providing those answers in the forthcoming era of big data.

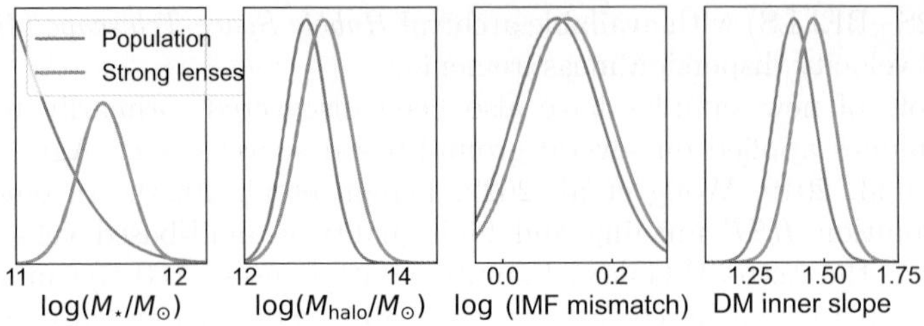

Figure 4. Impact of strong lensing selection on various properties of galaxies estimated by Sonnenfeld et al. (2023): from left to right, stellar mass $M_{\star}$, halo mass M_{halo}, IMF mismatch α, and dark matter (DM) inner slope. The IMF mismatch parameter in strong lensing galaxies is biased by 10% from the parent population, and the dark matter inner slope is biased by 5%.

Acknowledgements

AJS was supported by NASA through the NASA Hubble Fellowship grant HST-HF2-51492 awarded by the Space Telescope Science Institute, which is operated by the Association of Universities for Research in Astronomy, Inc., for NASA, under contract NAS5-26555.

References

Adam, A., Coogan, A., Malkin, N., Legin, R., Perreault-Levasseur, L., Hezaveh, Y., & Bengio, Y. 2022, Posterior samples of source galaxies in strong gravitational lenses with score-based priors. *arXiv e-prints*, arXiv:2211.03812.

Agnello, A., Grillo, C., Jones, T., Treu, T., Bonamigo, M., & Suyu, S. H. 2018, Discovery and first models of the quadruply lensed quasar sdss j1433+6007. *Monthly Notices of the Royal Astronomical Society*, 474(3), 3391–3396.

Akhazhanov, A., More, A., Amini, A., Hazlett, C., Treu, T., Birrer, S., Shajib, A., Liao, K., Lemon, C., Agnello, A., Nord, B., Aguena, M., Allam, S., Andrade-Oliveira, F., Annis, J., Brooks, D., Buckley-Geer, E., Burke, D. L., Carnero Rosell, A., Carrasco Kind, M., Carretero, J., Choi, A., Conselice, C., Costanzi, M., da Costa, L. N., Pereira, M. E. S., De Vicente, J., Desai, S., Dietrich, J. P., Doel, P., Everett, S., Ferrero, I., Finley, D. A., Flaugher, B., Frieman, J., García-Bellido, J., Gerdes, D. W., Gruen, D., Gruendl, R. A., Gschwend, J., Gutierrez, G., Hinton, S. R., Hollowood, D. L., Honscheid, K., James, D. J., Kim, A. G., Kuehn, K., Kuropatkin, N., Lahav, O., Lima, M., Lin, H., Maia, M. A. G., March, M., Menanteau, F., Miquel, R., Morgan, R., Palmese, A., Paz-Chinchón, F., Pieres, A., Plazas Malagón, A. A., Sanchez, E., Scarpine, V., Serrano, S., Sevilla-Noarbe, I., Smith, M., Soares-Santos, M., Suchyta, E., Swanson, M. E. C., Tarle, G., To, C., Varga, T. N., Weller, J., & (DES Collaboration) 2022, Finding quadruply imaged quasars with machine learning - I. Methods. *Monthly Notices of the Royal Astronomical Society*, 513(2), 2407–2421.

Auger, M. W., Treu, T., Bolton, A. S., Gavazzi, R., Koopmans, L. V. E., Marshall, P. J., Bundy, K., & Moustakas, L. A. 2009, The sloan lens acs survey. ix. colors, lensing, and stellar masses of early-type galaxies. *Astrophysical Journal*, 705(2), 1099–1115.

Biggio, L., Vernardos, G., Galan, A., Peel, A., & Courbin, F. 2023, Modeling lens potentials with continuous neural fields in galaxy-scale strong lenses. *Astronomy & Astrophysics*, 675, A125.

Birrer, S. & Amara, A. 2018, lenstronomy: Multi-purpose gravitational lens modelling software package. *Physics of the Dark Universe*, 22, 189–201.

Birrer, S., Shajib, A. J., Gilman, D., Galan, A., Aalbers, J., Millon, M., Morgan, R., Pagano, G., Park, J. W., Teodori, L., Tessore, N., Ueland, M., Vyvere, L. V. d., Wagner-Carena, S., Wempe, E., Yang, L., Ding, X., Schmidt, T., Sluse, D., Zhang, M., & Amara, A. 2021, lenstronomy ii: A gravitational lensing software ecosystem. *JOSS*, 6(62), 3283.

Bolton, A. S., Brownstein, J. R., Kochanek, C. S., Shu, Y., Schlegel, D. J., Eisenstein, D. J., Wake, D. A., Connolly, N., Maraston, C., Arneson, R. A., & Weaver, B. A. 2012, The BOSS Emission-Line Lens Survey. II. Investigating Mass-density Profile Evolution in the SLACS+BELLS Strong Gravitational Lens Sample. *Astrophysical Journal*, 757(1), 82.

Bolton, A. S., Burles, S., Koopmans, L. V. E., Treu, T., & Moustakas, L. A. 2006, The sloan lens acs survey. i. a large spectroscopically selected sample of massive early-type lens galaxies. *Astrophysical Journal*, 638, 703–724.

Cappellari, M. 2016, Structure and kinematics of early-type galaxies from integral field spectroscopy. *Annual Review of Astronomy and Astrophysics*, 54, 597–665.

Cappellari, M., McDermid, R. M., Alatalo, K., Blitz, L., Bois, M., Bournaud, F., Bureau, M., Crocker, A. F., Davies, R. L., Davis, T. A., de Zeeuw, P. T., Duc, P.-A., Emsellem, E., Khochfar, S., Krajnović, D., Kuntschner, H., Lablanche, P.-Y., Morganti, R., Naab, T., Oosterloo, T., Sarzi, M., Scott, N., Serra, P., Weijmans, A.-M., & Young, L. M. 2012, Systematic variation of the stellar initial mass function in early-type galaxies. *Nature*, 484(7395), 485–488.

Collett, T. E. 2015, The population of galaxy-galaxy strong lenses in forthcoming optical imaging surveys. *Astrophysical Journal*, 811, 20.

Delchambre, L., Krone-Martins, A., Wertz, O., Ducourant, C., Galluccio, L., Klüter, J., Mignard, F., Teixeira, R., Djorgovski, S. G., Stern, D., Graham, M. J., Surdej, J., Bastian, U., Wambsganss, J., Le Campion, J. F., & Slezak, E. 2019, Gaia GraL: Gaia DR2 Gravitational Lens Systems. III. A systematic blind search for new lensed systems. *Astronomy & Astrophysics*, 622, A165.

Dutton, A. A. & Treu, T. 2014, The bulge-halo conspiracy in massive elliptical galaxies: implications for the stellar initial mass function and halo response to baryonic processes. *Monthly Notices of the Royal Astronomical Society*, 438, 3594–3602.

Ertl, S., Schuldt, S., Suyu, S. H., Schmidt, T., Treu, T., Birrer, S., Shajib, A. J., & Sluse, D. 2023, TDCOSMO. X. Automated modeling of nine strongly lensed quasars and comparison between lens-modeling software. *Astronomy & Astrophysics*, 672, A2.

Etherington, A., Nightingale, J. W., Massey, R., Cao, X., Robertson, A., Amorisco, N. C., Amvrosiadis, A., Cole, S., Frenk, C. S., He, Q., Li, R., & Tam, S.-I. 2022, Automated galaxy-galaxy strong lens modelling: No lens left behind. *Monthly Notices of the Royal Astronomical Society*, 517(3), 3275–3302.

Ferreras, I., Saha, P., Leier, D., Courbin, F., & Falco, E. E. 2010, Constraining the low-mass end of the initial mass function with gravitational lensing. *Monthly Notices of the Royal Astronomical Society*, 409(1), L30–L34.

Hezaveh, Y. D., Levasseur, L. P., & Marshall, P. J. 2017, Fast automated analysis of strong gravitational lenses with convolutional neural networks. *Nature*, 548, 555–557.

Jacobs, C., Collett, T., Glazebrook, K., Buckley-Geer, E., Diehl, H. T., Lin, H., McCarthy, C., Qin, A. K., Odden, C., Caso Escudero, M., Dial, P., Yung, V. J., Gaitsch, S., Pellico, A., Lindgren, K. A., Abbott, T. M. C., Annis, J., Avila, S., Brooks, D., Burke, D. L., Carnero Rosell, A., Carrasco Kind, M., Carretero, J., da Costa, L. N., De Vicente, J., Fosalba, P., Frieman, J., García-Bellido, J., Gaztanaga, E., Goldstein, D. A., Gruen, D., Gruendl, R. A., Gschwend, J., Hollowood, D. L., Honscheid, K., Hoyle, B., James, D. J., Krause, E., Kuropatkin, N., Lahav, O., Lima, M., Maia, M. A. G., Marshall, J. L., Miquel, R., Plazas, A. A., Roodman, A., Sanchez, E., Scarpine, V., Serrano, S., Sevilla-Noarbe, I., Smith, M., Sobreira, F., Suchyta, E., Swanson, M. E. C., Tarle, G., Vikram, V., Walker, A. R., Zhang, Y., & DES Collaboration 2019,a An extended catalog of galaxy-galaxy strong gravitational lenses discovered in des using convolutional neural networks. *Astrophysical Journal Supplement*, 243a(1), 17.

Jacobs, C., Collett, T., Glazebrook, K., McCarthy, C., Qin, A. K., Abbott, T. M. C., Abdalla, F. B., Annis, J., Avila, S., Bechtol, K., Bertin, E., Brooks, D., Buckley-Geer, E., Burke, D. L., Carnero Rosell, A., Carrasco Kind, M., Carretero, J., da Costa, L. N., Davis, C., De Vicente, J., Desai, S., Diehl, H. T., Doel, P., Eifler, T. F., Flaugher, B., Frieman, J., García-Bellido, J., Gaztanaga, E., Gerdes, D. W., Goldstein, D. A., Gruen, D., Gruendl,

R. A., Gschwend, J., Gutierrez, G., Hartley, W. G., Hollowood, D. L., Honscheid, K., Hoyle, B., James, D. J., Kuehn, K., Kuropatkin, N., Lahav, O., Li, T. S., Lima, M., Lin, H., Maia, M. A. G., Martini, P., Miller, C. J., Miquel, R., Nord, B., Plazas, A. A., Sanchez, E., Scarpine, V., Schubnell, M., Serrano, S., Sevilla-Noarbe, I., Smith, M., Soares-Santos, M., Sobreira, F., Suchyta, E., Swanson, M. E. C., Tarle, G., Vikram, V., Walker, A. R., Zhang, Y., Zuntz, J., & DES Collaboration 2019,b Finding high-redshift strong lenses in des using convolutional neural networks. *Monthly Notices of the Royal Astronomical Society*, 484b(4), 5330–5349.

La Barbera, F., Ferreras, I., Vazdekis, A., de la Rosa, I. G., de Carvalho, R. R., Trevisan, M., Falcón-Barroso, J., & Ricciardelli, E. 2013, Spider viii - constraints on the stellar initial mass function of early-type galaxies from a variety of spectral features. *Monthly Notices of the Royal Astronomical Society*, 433(4), 3017–3047.

Lemon, C., Anguita, T., Auger-Williams, M. W., Courbin, F., Galan, A., McMahon, R., Neira, F., Oguri, M., Schechter, P., Shajib, A., Treu, T., Agnello, A., & Spiniello, C. 2023, Gravitationally lensed quasars in Gaia - IV. 150 new lenses, quasar pairs, and projected quasars. *Monthly Notices of the Royal Astronomical Society*, 520(3), 3305–3328.

Mishra-Sharma, S. & Yang, G. Strong Lensing Source Reconstruction Using Continuous Neural Fields. In *Machine Learning for Astrophysics* 2022, 34.

Morningstar, W. R., Perreault Levasseur, L., Hezaveh, Y. D., Blandford, R., Marshall, P., Putzky, P., Rueter, T. D., Wechsler, R., & Welling, M. 2019, Data-driven Reconstruction of Gravitationally Lensed Galaxies Using Recurrent Inference Machines. *Astrophysical Journal*, 883(1), 14.

Mukherjee, S., Koopmans, L. V. E., Metcalf, R. B., Tortora, C., Schaller, M., Schaye, J., Vernardos, G., & Bellagamba, F. 2021, Seagle - ii. constraints on feedback models in galaxy formation from massive early-type strong-lens galaxies. *Monthly Notices of the Royal Astronomical Society*, 504, 3455–3477.

Myers, S. T., Fassnacht, C. D., Djorgovski, S. G., Blandford, R. D., Matthews, K., Neugebauer, G., Pearson, T. J., Readhead, A. C. S., Smith, J. D., Thompson, D. J., Womble, D. S., Browne, I. W. A., Wilkinson, P. N., Nair, S., Jackson, N., Snellen, I. A. G., Miley, G. K., de Bruyn, A. G., & Schilizzi, R. T. 1995, 1608+656: A Quadruple-Lens System Found in the CLASS Gravitational Lens Survey. *Astrophysical Journal Letters*, 447, L5.

Navarro, J. F., Frenk, C. S., & White, S. D. M. 1996, The structure of cold dark matter halos. *Astrophysical Journal*, 462, 563.

Navarro, J. F., Frenk, C. S., & White, S. D. M. 1997, A universal density profile from hierarchical clustering. *Astrophysical Journal*, 490, 493–508.

Nightingale, J., Hayes, R., Kelly, A., Amvrosiadis, A., Etherington, A., He, Q., Li, N., Cao, X., Frawley, J., Cole, S., Enia, A., Frenk, C., Harvey, D., Li, R., Massey, R., Negrello, M., & Robertson, A. 2021, PyAutoLens: Open-Source Strong Gravitational Lensing. *The Journal of Open Source Software*, 6(58), 2825.

Oguri, M. & Marshall, P. J. 2010, Gravitationally lensed quasars and supernovae in future wide-field optical imaging surveys. *Monthly Notices of the Royal Astronomical Society*, 405, 2579–2593.

Oldham, L. J. & Auger, M. W. 2018, Dark matter contraction and stellar-mass-to-light ratio gradients in massive early-type galaxies. *Monthly Notices of the Royal Astronomical Society*, 476(1), 133–150.

Poh, J., Samudre, A., Ćiprijanović, A., Nord, B., Khullar, G., Tanoglidis, D., & Frieman, J. A. 2022, Strong Lensing Parameter Estimation on Ground-Based Imaging Data Using Simulation-Based Inference. *arXiv e-prints*, arXiv:2211.05836.

Posacki, S., Cappellari, M., Treu, T., Pellegrini, S., & Ciotti, L. 2015, The stellar initial mass function of early-type galaxies from low to high stellar velocity dispersion: homogeneous analysis of atlas^{3d} and sloan lens acs galaxies. *Monthly Notices of the Royal Astronomical Society*, 446(1), 493–509.

Remus, R.-S., Dolag, K., Naab, T., Burkert, A., Hirschmann, M., Hoffmann, T. L., & Johansson, P. H. 2017, The co-evolution of total density profiles and central dark matter fractions in simulated early-type galaxies. *Monthly Notices of the Royal Astronomical Society*, 464(3), 3742–3756.

Ruff, A. J., Gavazzi, R., Marshall, P. J., Treu, T., Auger, M. W., & Brault, F. 2011, The SL2S Galaxy-scale Lens Sample. II. Cosmic Evolution of Dark and Luminous Mass in Early-type Galaxies. *Astrophysical Journal*, 727(2), 96.

Schmidt, T., Treu, T., Birrer, S., Shajib, A. J., Lemon, C., Millon, M., Sluse, D., Agnello, A., Anguita, T., Auger-Williams, M. W., McMahon, R. G., Motta, V., Schechter, P., Spiniello, C., Kayo, I., Courbin, F., Ertl, S., Fassnacht, C. D., Frieman, J. A., More, A., Schuldt, S., Suyu, S. H., Aguena, M., Andrade-Oliveira, F., Annis, J., Bacon, D., Bertin, E., Brooks, D., Burke, D. L., Carnero Rosell, A., Carrasco Kind, M., Carretero, J., Conselice, C., Costanzi, M., da Costa, L. N., Pereira, M. E. S., De Vicente, J., Desai, S., Doel, P., Everett, S., Ferrero, I., Friedel, D., García-Bellido, J., Gaztanaga, E., Gruen, D., Gruendl, R. A., Gschwend, J., Gutierrez, G., Hinton, S. R., Hollowood, D. L., Honscheid, K., James, D. J., Kuehn, K., Lahav, O., Menanteau, F., Miquel, R., Palmese, A., Paz-Chinchón, F., Pieres, A., Plazas Malagón, A. A., Prat, J., Rodriguez-Monroy, M., Romer, A. K., Sanchez, E., Scarpine, V., Sevilla-Noarbe, I., Smith, M., Suchyta, E., Tarle, G., To, C., & Varga, T. N. 2022, Strides: Automated uniform models for 30 quadruply imaged quasars. *arXiv e-prints*.

Schuldt, S., Suyu, S. H., Meinhardt, T., Leal-Taixé, L., Cañameras, R., Taubenberger, S., & Halkola, A. 2021, HOLISMOKES. IV. Efficient mass modeling of strong lenses through deep learning. *Astronomy & Astrophysics*, 646, A126.

Shajib, A. J., Birrer, S., Treu, T., Agnello, A., Buckley-Geer, E. J., Chan, J. H. H., Christensen, L., Lemon, C., Lin, H., Millon, M., Poh, J., Rusu, C. E., Sluse, D., Spiniello, C., Chen, G. C. F., Collett, T., Courbin, F., Fassnacht, C. D., Frieman, J., Galan, A., Gilman, D., More, A., Anguita, T., Auger, M. W., Bonvin, V., McMahon, R., Meylan, G., Wong, K. C., Abbott, T. M. C., Annis, J., Avila, S., Bechtol, K., Brooks, D., Brout, D., Burke, D. L., Carnero Rosell, A., Carrasco Kind, M., Carretero, J., Castander, F. J., Costanzi, M., da Costa, L. N., De Vicente, J., Desai, S., Dietrich, J. P., Doel, P., Drlica-Wagner, A., Evrard, A. E., Finley, D. A., Flaugher, B., Fosalba, P., García-Bellido, J., Gerdes, D. W., Gruen, D., Gruendl, R. A., Gschwend, J., Gutierrez, G., Hollowood, D. L., Honscheid, K., Huterer, D., James, D. J., Jeltema, T., Krause, E., Kuropatkin, N., Li, T. S., Lima, M., MacCrann, N., Maia, M. A. G., Marshall, J. L., Melchior, P., Miquel, R., Ogando, R. L. C., Palmese, A., Paz-Chinchón, F., Plazas, A. A., Romer, A. K., Roodman, A., Sako, M., Sanchez, E., Santiago, B., Scarpine, V., Schubnell, M., Scolnic, D., Serrano, S., Sevilla-Noarbe, I., Smith, M., Soares-Santos, M., Suchyta, E., Tarle, G., Thomas, D., Walker, A. R., & Zhang, Y. 2020, Strides: a 3.9 per cent measurement of the hubble constant from the strong lens system des j0408-5354. *Monthly Notices of the Royal Astronomical Society*, 494(4), 6072–6102.

Shajib, A. J., Birrer, S., Treu, T., Auger, M. W., Agnello, A., Anguita, T., Buckley-Geer, E. J., Chan, J. H. H., Collett, T. E., Courbin, F., Fassnacht, C. D., Frieman, J., Kayo, I., Lemon, C., Lin, H., Marshall, P. J., McMahon, R., More, A., Morgan, N. D., Motta, V., Oguri, M., Ostrovski, F., Rusu, C. E., Schechter, P. L., Shanks, T., Suyu, S. H., Meylan, G., Abbott, T. M. C., Allam, S., Annis, J., Avila, S., Bertin, E., Brooks, D., Carnero Rosell, A., Carrasco Kind, M., Carretero, J., Cunha, C. E., da Costa, L. N., De Vicente, J., Desai, S., Doel, P., Flaugher, B., Fosalba, P., García-Bellido, J., Gerdes, D. W., Gruen, D., Gruendl, R. A., Gutierrez, G., Hartley, W. G., Hollowood, D. L., Hoyle, B., James, D. J., Kuehn, K., Kuropatkin, N., Lahav, O., Lima, M., Maia, M. A. G., March, M., Marshall, J. L., Melchior, P., Menanteau, F., Miquel, R., Plazas, A. A., Sanchez, E., Scarpine, V., Sevilla-Noarbe, I., Smith, M., Soares-Santos, M., Sobreira, F., Suchyta, E., Swanson, M. E. C., Tarle, G., & Walker, A. R. 2019, Is every strong lens model unhappy in its own way? uniform modelling of a sample of 13 quadruply+ imaged quasars. *Monthly Notices of the Royal Astronomical Society*, 483, 5649–5671.

Shajib, A. J., Treu, T., Birrer, S., & Sonnenfeld, A. 2021, Dark matter haloes of massive elliptical galaxies at z $\sim$ 0.2 are well described by the navarro-frenk-white profile. *MNRAS*, 503(2), 2380–2405.

Shajib, A. J., Vernardos, G., Collett, T. E., Motta, V., Sluse, D., Williams, L. L. R., Saha, P., Birrer, S., Spiniello, C., & Treu, T. 2022, Strong lensing by galaxies.

Smith, R. J., Lucey, J. R., & Conroy, C. 2015, The sinfoni nearby elliptical lens locator survey: discovery of two new low-redshift strong lenses and implications for the initial mass function in giant early-type galaxies. *Monthly Notices of the Royal Astronomical Society*, 449, 3441–3457.

Sonnenfeld, A., Chan, J. H. H., Shu, Y., More, A., Oguri, M., Suyu, S. H., Wong, K. C., Lee, C.-H., Coupon, J., Yonehara, A., Bolton, A. S., Jaelani, A. T., Tanaka, M., Miyazaki, S., & Komiyama, Y. 2018, Survey of gravitationally-lensed objects in hsc imaging (sugohi). i. automatic search for galaxy-scale strong lenses. *Publications of the Astronomical Society of Japan*, 70, S29.

Sonnenfeld, A., Gavazzi, R., Suyu, S. H., Treu, T., & Marshall, P. J. 2013, The sl2s galaxy-scale lens sample. iii. lens models, surface photometry, and stellar masses for the final sample. *Astrophysical Journal*, 777, 97.

Sonnenfeld, A., Li, S.-S., Despali, G., Shajib, A. J., & Taylor, E. N. 2023, Strong lensing selection effects. *arXiv e-prints*, arXiv:2301.13230.

Sonnenfeld, A., Treu, T., Gavazzi, R., Marshall, P. J., Auger, M. W., Suyu, S. H., Koopmans, L. V. E., & Bolton, A. S. 2012, Evidence for dark matter contraction and a salpeter initial mass function in a massive early-type galaxy. *Astrophysical Journal*, 752, 163.

Sonnenfeld, A., Treu, T., Marshall, P. J., Suyu, S. H., Gavazzi, R., Auger, M. W., & Nipoti, C. 2015, The sl2s galaxy-scale lens sample. v. dark matter halos and stellar imf of massive early-type galaxies out to redshift 0.8. *Astrophysical Journal*, 800, 94.

Sonnenfeld, A., Wang, W., & Bahcall, N. 2019, Hyper suprime-cam view of the cmass galaxy sample. halo mass as a function of stellar mass, size, and sérsic index. *Astronomy & Astrophysics*, 622, A30.

Spiniello, C., Koopmans, L. V. E., Trager, S. C., Czoske, O., & Treu, T. 2011, The x-shooter lens survey - i. dark matter domination and a salpeter-type initial mass function in a massive early-type galaxy. *Monthly Notices of the Royal Astronomical Society*, 417(4), 3000–3009.

Spiniello, C., Trager, S., Koopmans, L. V. E., & Conroy, C. 2014, The stellar imf in early-type galaxies from a non-degenerate set of optical line indices. *Monthly Notices of the Royal Astronomical Society*, 438(2), 1483–1499.

Tran, K.-V. H., Harshan, A., Glazebrook, K., Keerthi Vasan, G. C., Jones, T., Jacobs, C., Kacprzak, G. G., Barone, T. M., Collett, T. E., Gupta, A., Henderson, A., Kewley, L. J., Lopez, S., Nanayakkara, T., Sanders, R. L., & Sweet, S. M. 2022, The agel survey: Spectroscopic confirmation of strong gravitational lenses in the des and decals fields selected using convolutional neural networks. *The Astronomical Journal*, 164, 148.

Treu, T. 2010, Strong lensing by galaxies. *Annual Review of Astronomy and Astrophysics*, 48, 87–125.

Treu, T., Auger, M. W., Koopmans, L. V. E., Gavazzi, R., Marshall, P. J., & Bolton, A. S. 2010, The initial mass function of early-type galaxies. *Astrophysical Journal*, 709(2), 1195–1202.

Wang, Y., Vogelsberger, M., Xu, D., Mao, S., Springel, V., Li, H., Barnes, D., Hernquist, L., Pillepich, A., Marinacci, F., Pakmor, R., Weinberger, R., & Torrey, P. 2020, Early-type galaxy density profiles from illustristng - i. galaxy correlations and the impact of baryons. *Monthly Notices of the Royal Astronomical Society*, 491(4), 5188–5215.

Wong, K. C., Chan, J. H. H., Chao, D. C. Y., Jaelani, A. T., Kayo, I., Lee, C.-H., More, A., & Oguri, M. 2022, Survey of Gravitationally lensed objects in HSC Imaging (SuGOHI). VIII. New galaxy-scale lenses from the HSC SSP. *Publications of the Astronomical Society of Japan*, 74(5), 1209–1219.

Xu, D., Springel, V., Sluse, D., Schneider, P., Sonnenfeld, A., Nelson, D., Vogelsberger, M., & Hernquist, L. 2017, The inner structure of early-type galaxies in the illustris simulation. *Monthly Notices of the Royal Astronomical Society*, 469(2), 1824–1848.

Strong Gravitational Lensing in the Era of Big Data
Proceedings IAU Symposium No. 381, 2024
H. Stacey, A. Sonnenfeld & C. Grillo, eds.
doi:10.1017/S1743921323003691

The "External" Shears In Strong Lens Models

James W. Nightingale, Amy Etherington and Richard Massey

Department of Physics, Centre for Extragalactic Astronomy, Durham University, South Rd, Durham, DH1 3LE, UK. email: james.w.nightingale@durham.ac.uk

Abstract. The distribution of mass in galaxy-scale strong gravitational lenses is often modelled as an elliptical power law plus 'external shear', which notionally accounts for line-of-sight galaxies and cosmic shear. We argue that it does not, using three lines of evidence from the analysis of 54 galaxy-scale strong lenses: (i) strong lensing external shears do not correlate with weak lensing; (ii) the measured shear magnitudes in strong lenses (which are field galaxies) are too large (exceeding 0.05) for their environment and; (iii) the external shear position angle preferentially aligns or anti-aligns with the mass model position angle, indicating an internal origin. We argue the measured strong lensing shears are therefore systematically accounting for missing complexity in the canonical elliptical power-law mass model. If we can introduce this complexity into our lens models, this will further lensing studies of galaxy formation, dark matter and Cosmology.

Keywords. Strong lensing, galaxies

Strong gravitational lens modeling necessitates the assumption of a mass model for the lens galaxy, in order to trace light from the image-plane to the source-plane. Typically, a power-law mass distribution, often with a fixed slope for isothermality, is assumed. This represents the total mass of the lens galaxy, including stars and dark matter.

In addition to this mass model, strong lens models frequently incorporate an external shear term. Conventionally, this term is thought to account for the lensing effects of nearby line-of-sight galaxies surrounding the lens of interest. It is a common refrain in lens modeling literature to acknowledge the inclusion of this external shear term, as exemplified by the quote from a paper by the author of this section (Nightingale 2019), "Both of the mass models above include an external shear term, which accounts for the contribution of line-of-sight structures to the strong lensing signal."

However, a closer examination of the literature over the past few decades suggests that the external shear in strong lens models might not actually represent the influence of nearby line-of-sight galaxies. This doubt arises from several findings, including the alignment of shear position angles with the lens mass model position angles (Witt 1997), a lack of correlation between shears derived from lens modeling and environmental or weak lensing studies (Wong 2011), and inconsistencies in the lens models of observed quadruply imaged quasars (Gomer 2021). Notably, (Keeton 1997), a frequently cited source when introducing the external shear in a lens model, highlights themselves that "the secondary shear must also be dominated by the primary lens galaxy rather than external perturbations," casting doubt on the external shear's origin. However, it's crucial to acknowledge that these studies often work with limited sample sizes, preventing conclusive determinations.

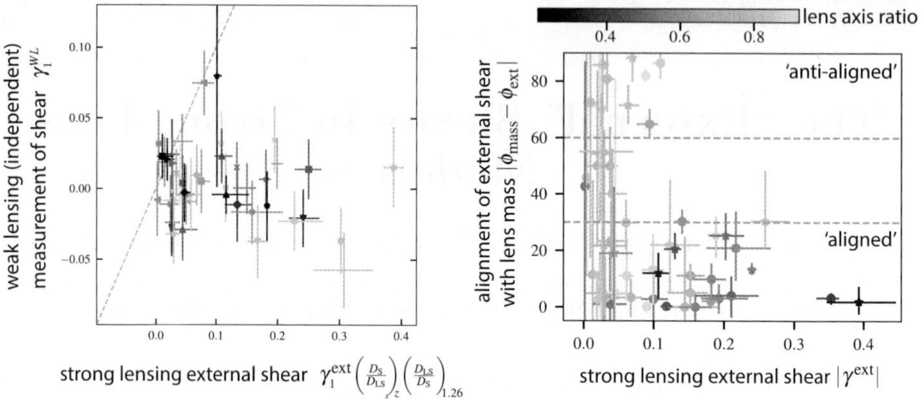

Figure 1. Left Panel: Values of the shear along the lines of sight to 39 galaxy-galaxy lenses, independently measured using strong lensing 'external' shear γ^{SL} and weak lensing γ^{WL}. Strong lensing shears do not correlate with weak lensing shears and go to magnitudes in excess of 0.05. **Right Panel:** Relative orientation of the strong lensing 'external' shear and the major axis of the lens mass, in 54 galaxy-galaxy lenses. In both cases, most of the shears are suspiciously aligned ($\phi_{\rm mass}^{\rm PL+ext} - \phi_{\rm ext}^{\rm PL+ext} \leq 30°$) or anti-aligned ($\phi_{\rm mass}^{\rm PL+ext} - \phi_{\rm ext}^{\rm PL+ext} \geq 60°$) with the lens mass distribution. Points are coloured by the best-fit axis ratio of the lens mass distribution. Highly elliptical lenses often lead to high values of $\gamma_{\rm ext}$.

We presented the results of (Etherington 2023), offering compelling evidence that the shears inferred in strong lens models are predominantly of internal origin. The findings are based on the comprehensive analysis of (Etherington 2022), who performed uniform modeling of 54 strong lenses from the SLACS (Bolton 2008) and BELLS-GALLERY (Shu 2016) samples, using the open-source lens modeling software `PyAutoLens` (Nightingale 2015; Nightingale 2018; Nightingale 2021). The lens modeling assumes an elliptical power-law mass model with external shear, and is supplemented with weak lensing analysis of the SLACS sample that are compared with the strong lens models.

In Figure 1, we present three key pieces of evidence suggesting that the dominant contributor to the shears in strong lens models is not an external signal from nearby line-of-sight galaxies but something else entirely:

• **Lack of Correlation Between Strong and Weak Lensing Shears:** We begin by examining the absence of correlation between strong and weak lensing shears. Weak lensing measurements, which capture shear on larger environmental scales, are expected to closely resemble external shears. The left panel of Figure 1 compares the strong lensing shear $\gamma_1^{\rm SL}$ with the weak lensing shear $\gamma_1^{\rm WL}$. This reveals no such correlation between strong lensing shear ($\gamma_1^{\rm SL}$) and weak lensing shear ($\gamma_1^{\rm WL}$). This discrepancy implies that the strong lens shear is measuring an entirely different phenomenon.

• **Excessive Shear Magnitudes:** We then delve into the issue of shear magnitudes. Strong lenses typically exist in field environments where shear magnitudes should never exceed 0.05, as shear of such magnitude is usually only encountered in galaxy clusters. The x-axis of the right panels of Figure 1 show $\gamma^{\rm ext}$, the inferred shear magnitudes of the strong lens models fitted to all 54 lenses. It shows that approximately half of the lenses in the sample exhibit shear magnitudes exceeding 0.05, with over a quarter exceeding 0.1. These values are far too large for galaxies inhabiting field environments.

• **Alignment of Strong Lensing Shear Position Angles with Mass Model:** The final piece of evidence centers on the alignment of strong lensing shear position angles with the mass model's position angles. If the shear were indeed external, we would expect no correlation with internal mass model parameters. The y-axis of the right panel shows the of Figure 1 shows the absolute difference of the position angle in degrees of the power-law mass model and strong lens model external shear. A value of $0°$ indicates the two are aligned, which occurs in ~ 13 lenses, another ~ 5 lenses show anti-alignment with a value of $90°$, and the overall lens samples clusters towards these values. The strong lens model external shear is tightly coupled to the lens mass models, indicating an internal origin.

Collectively, these lines of evidence suggest that the primary contributor to the external shears measured in strong lens models is not a genuine external shear from nearby line-of-sight galaxies. Instead, it appears to be driven by other factors, which are yet to be definitively identified. It is expected that a genuine external shear signal from the surrounding environment is present in every strong lens, it is simply sub-dominant to these other factors.

This raises the fundamental question of what the dominant contributor to the measured strong lensing shears might be. (Etherington 2023) argue that the simplicity of the elliptical power-law mass model commonly assumed for lens galaxies is inadequate. This model assumes that the mass distribution can be described by isodensity contours with a single axis ratio and position angle, failing to account for complexities like radial twists in the mass profile and boxiness/diskiness in iso-density contours. Recent simulations and studies support this view, showing that complex mass features, such as boxiness, diskiness, and mass twists, can affect shear measurements in lens models (Cao 2020; VanDeVyvere 2022a; VanDeVyvere 2022b).

At present, we lack a definitive explanation for why strong lens models yield these apparently "external shears." Emphasis must now shift towards understanding the aspects of real lens galaxy's mass distributions that the power-law mass model fails to capture. Looking ahead, this pursuit serves two vital purposes.

Firstly, these "external shears" underlie the lens models used in critical areas of strong lens research, such as dark matter substructure (Vegetti 2014) and Hubble constant measurements (Suyu 2016). The absence of a clear understanding of what these shears represent could introduce unexpected systematics into analyses, with it being established that lens models can trade-off different forms of complex in ways that may bias cosmological measurements (Kochanek 2021) (see also the mass sheet transformation (Schneider 2014)). Ensuring that our models accurately represent the physical systems they are meant to describe is essential for robust analyses, and this is evidently not yet the case for strong lens models.

Secondly, despite the current uncertainty regarding the origin of "external shears," they represent a detected signal worthy of further scientific investigation. We anticipate that this signal results from a combination of complex mass structures within the lens, such as boxiness, diskiness, mass twists, and centroid offsets. Once we decipher this signal, it could offer profound insights into the mass structure of high-redshift galaxies, enriching our understanding of galaxy formation and evolution models. Additionally, a true external shear signal ought to be present in every strong lens observation, which contains cosmologically valuable information, particularly when combined with wide-field weak lensing cosmological surveys (Birrer 2017; Hogg 2022). By comprehending the source of the "external shear," we can unlock the potential to measure the true external shear in each lens, thereby advancing cosmological endeavors.

In conclusion, we implore authors of strong lens papers to refrain from characterizing the external shear in their models as merely "accounting for nearby line-of-sight galaxies." Instead, we encourage them to acknowledge its role in shaping the internal mass complexity of the lens. This shift in perspective will not only rectify a misconception prevalent throughout the literature but also guide us toward a deeper comprehension of the underlying physical nature of the lens galaxies we scrutinize.

References

Bolton A. S., et al., 2008, ApJ, 682, 964
Birrer S., et al., 2017, Journal of Cosmology and Astroparticle Physics, 2017
Cao X. et al., 2021, RAA, 22, 30 pp
Etherington A. et al., 2023b, arXiv:2301.05244
Etherington A. et al., 2022, MNRAS, 517, 3275
Gomer M. R., Williams L. L. R., 2021, MNRAS, 504, 1340
Hogg N. et al., 2023, MNRAS, 520, 5982
Keeton C. R., Kochanek C. S., Seljak U., 1997, ApJ, 482, 604
Kochanek C. S., 2021, MNRAS, 501, 5021
Nightingale J. W., Dye S., 2015, MNRAS, 452, 2940
Nightingale J., Dye S., Massey R., 2018, MNRAS, 47, 1
Nightingale J. W., et al., 2019, MNRAS
Nightingale J. et al., 2021, J. Open Source Softw., 6, 282
Schneider P., 2014, arXiv:1409.0015
Shu Y. et al., 2016, ApJ, 824, 86
Suyu et al., 2016, MNRAS, 468, 2590
Van De Vyvere L., et al., 2022a, A&A, 659, 1
Van De Vyvere L., et al., 2022b, A&A, 179, 1
Vegetti S., et al., 2014, MNRAS, 442, 2017
Witt H. J., Mao S., 1997, MNRAS, 291, 211
Wong K. C., et al., 2011, ApJ, 726

Strong Gravitational Lensing in the Era of Big Data
Proceedings IAU Symposium No. 381, 2024
H. Stacey, A. Sonnenfeld & C. Grillo, eds.
doi:10.1017/S174392132300371X

Strong lensing by edge-on galaxies in UNIONS

J. A. Acevedo Barroso[1], B. Clément[2], F. Courbin[1], R. Gavazzi[2,3], C. Lemon[1], K. Rojas[4] and E. Savary[1,5]

[1]Institute of Physics, Laboratory of Astrophysics, Ecole Polytechnique Fédérale de Lausanne (EPFL), Observatoire de Sauverny, 1290 Versoix, Switzerland.
email: javier.acevedobarroso@epfl.ch

[2]Aix-Marseille Université, CNRS, CNES, LAM, Marseille, France

[3]Institut d'Astrophysique de Paris, UMR 7095, CNRS, and Sorbonne Université, 98 bis boulevard Arago, 75014 Paris, France

[4]Institute of Cosmology and Gravitation, University of Portsmouth, Burnaby Rd, Portsmouth PO1 3FX, UK

[5]Departement of Radiology, Lausanne University Hospital and University of Lausanne, Lausanne, Switzerland

Abstract. Current searches for galaxy-scale strong lenses focus on massive Luminous Red Galaxies but tend to overlook late-type lenses, in part because of their smaller Einstein radii. We take advantage of the superb seeing of the UNIONS survey in the *r*-band to perform an imaging search for edge-on late-type lenses. We use Convolutional Neural Networks trained with simulated observations composed of images of real galaxies from UNIONS and real sources from HST. Using 3600 square degrees of the survey we test ∼7 million galaxies and find 56 systems with obvious signs of lensing. In addition, we empirically estimate the true prevalence of lenses in UNIONS by visually inspecting 120,000 randomly chosen images in the survey. We find that the number of edge-on lenses we discover with CNNs is compatible with these estimates.

Keywords. Gravitational lensing: strong, galaxies: spiral, methods: data analysis, surveys.

1. Introduction

Machine Learning methods, applied to wide-field imaging surveys, have led to a revolution in the field of galaxy-scale strong lensing. Just in the past five years, the number of known strong lenses has increased by one order of magnitude, with some of the most recent lens-finding works reporting, on their own, tens or even hundreds of reliable candidates (e.g. Cañameras et al. 2020; Li et al. 2021; Rojas et al. 2022; Savary et al. 2022). However, the new lenses discovered are vastly dominated by early-type galaxies such as Luminous Red Galaxies (LRG) and late-type lenses remain a minority. This can be partly explained by the smaller lensing cross section of late-type galaxies, making them less likely than LRGs of producing prominent lensing features. In addition, current lens-finding teams tend to deliberately bias their samples towards LRG lenses, as indeed this is where chances of finding efficient lenses are the best. As a result, using strong lensing as a tool to study galaxy formation and evolution remains limited to a fairly narrow range of galaxy morphologies, leaving late-type galaxies underrepresented.

The lensing cross-section of late-type galaxies peaks when they are seen edge-on, hence maximizing the projected surface mass density along the line of sight. And indeed, most examples of late-type lenses do have edge-on deflectors. This turns out to be very convenient for spectroscopic follow-up aiming at measuring the stellar mass and the rotation curve of the lens. The combination of kinematics with lens modelling allows us to separate the stellar and dark mass distributions as well as the IMF of the lens (e.g. Ferreras et al. 2010) and to break known model degeneracies such as the disc-bulge degeneracy or the dark matter-baryons degeneracy (e.g. Dutton et al. 2011). This technique, however, has not been systematically exploited in the last decade due to the low number of known edge-on disc lenses.

The largest current sample of late-type lenses resides in the *Sloan WFC Edge-on Late-type Lens Survey* (SWELLS, Treu et al. 2011). The SWELLS team looked for SDSS (Sloan Digital Sky Survey) spectra that contained more than one set of lines at different redshifts and then obtained high-resolution imaging using the *Hubble Space Telescope* (HST). They confirmed and studied 24 edge-on lenses, and studied in detail 20 of them (Brewer et al. 2012). By contrast, the largest search using wide-field optical imaging was carried out by Sygnet et al. (2010), finding 11 good candidates in the 150 deg^2 wide part of the Canada-France Hawaii Telescope (CFHT) Legacy Survey. This study did not use Machine Learning. To date, there has been no systematic and targeted search for edge-on lenses in wide-field imaging surveys using Machine Learning. Most new edge-on lens candidates were in fact discovered in the course of LRG lens searches, and thus, are found only in tiny proportions. For example, out of 294 candidates in Cañameras et al. (2021), only 13 have edge-on late-type galaxies as main deflectors.

In this work, we use a *Convolutional Neural Network* (CNN) to do a specific search for late-type lens galaxies, using data-driven imaging simulations composed exclusively of edge-on lenses for the positive class and of a random selection of galaxies of all types for the negative class. The observations used here are part of the *Canada-France Infrared Survey* (CFIS), now part of the *Ultraviolet Near Infrared Optical Northern Survey* (UNIONS). Our search is carried in a single photometric band (r-band) under exquisite seeing conditions. As such, our work can be seen as exploratory work for the single-band VIS *Euclid* data.

2. Data

The Ultraviolet Near Infrared Optical Northern Survey (UNIONS) is a collaboration of wide field imaging surveys of the northern hemisphere. UNIONS consists of the Canada-France Imaging Survey (CFIS), conducted at the 3.6-meter CFHT on Maunakea, members of the Pan-STARRS team, and the Wide Imaging with Subaru HyperSuprime-Cam of the Euclid Sky (WISHES) team. CFHT/CFIS is obtaining deep u and r bands; Pan-STARRS is obtaining deep i and moderate-deep z band imaging, and Subaru is obtaining deep z-band imaging through WISHES and g-band imaging through the Waterloo-Hawaii IfA g-band Survey (WHIGS). These independent efforts are directed, in part, to securing optical imaging to complement the Euclid space mission, although UNIONS is a separate collaboration aimed at maximizing the science return of these large and deep surveys of the northern skies. We focus our search in the CFIS r-band data taking advantage of the excellent average seeing of 0.65 arcsec. The observations were taken with the wide-field imager MegaCam and reduced using the `MegaPipe` image processing pipeline (Gwyn 2008). Catalogs were produced using Sextractor (Bertin and Arnouts 1996, 2010).

We used observations covering 3624 deg^2 and applied a magnitude cut of $17 < r < 20.5$. We did not apply any ellipticity cut because the lensing features, when present, reduce the measured ellipticity of the whole system. This left us with 6,978,977 extended sources to mine for lenses. Our stamp size was 12.3 arcsec^2 (66 pixels on a side) centered on the source, along with a local model of the PSF from PSFEx (Bertin 2011, 2013) and a weight map.

To train the CNN we simulated a data set mocking CFIS r-band observations of edge-on lenses. We used real CFIS images for the deflector galaxies and HST imaging for the background sources. To select edge-on spirals we applied an ellipticity cut ($0.6 < \frac{a^2-b^2}{a^2+b^2} < 0.92$) and a magnitude cut. We limited the ellipticity cut to 0.92 in order to avoid diffraction spikes, dead columns and other spurious detections. This selection was matched with SDSS DR18 (Almeida et al. 2023) to obtain photometric redshifts. We kept only the galaxies that had good estimates of the photo-z error. In total, there were 982,721 potential deflectors for the mock edge-on lenses. To simulate the sources we used 50,889 r-band images from the catalog compiled by Cañameras et al. (2020), which combines HST/ACS F814W high-resolution images with color information from the *Hyper-Supreme Camera* (HSC) to produce high-resolution images mimicking the HSC observed colors.

3. Methodology

We used a supervised approach that relies on showing correctly-classified examples to the CNN, and minimizing the error between the score and the correct classification (1 for lenses, 0 otherwise). Given the low number of known edge-on lenses, we are forced to create a data set of mock CFIS observations of edge-on lenses. The lens-finding procedure has three main steps:

(1) Create a training data set by mocking CFIS observations of edge-on lenses.
(2) Train and deploy a CNN to select good edge-on lens candidates.
(3) Visually inspect the CNN's candidates to clean the final catalog from false-positives.

To make a mock lens we start with a potential deflector and assume a mass-to-light ratio for it. Then, we create a mass model based on the light profile of the galaxy plus a dark matter halo. Finally, we lens an HST source and add it. This was done using the `Lenstronomy` package (Birrer and Amara 2018; Birrer et al. 2021), where the luminous mass is represented with a Chameleon profile (see appendix of Dutton et al. 2011) and where the dark matter is a Navarro-Frenk-White profile (Navarro et al. 1996). The final training data set consisted of 99,547 mock lenses and an equal number of visually inspected non-lenses, of which 80% was used as training data, 10% as validation data, and 10% as test data.

For the classification we reimplemented the CNN `CMU DeepLens` (Lanusse et al. 2018) in Keras† (Chollet and others 2018) and trained it for 160 epochs using the standard binary cross-entropy as loss function and a Stochastic Gradient Descent optimizer. We used a learning rate, momentum and batch size of 0.001, 0.9 and 512 respectively. We consider objects with score larger than 0.5 as candidates for visual inspection. After training, the accuracy, precision and recall of the CNN on the test data set were 99.33%, 99.30% and 99.34%, respectively.

The visual inspection was done by J. A. B., B. C., and F. C. using a visualization tool derived from the one used in Savary et al. (2022)‡. The inspection was done in three steps: first, each expert inspected all of the candidates by *mosaics* of 10×10 stamps per page, selecting any object that shows any possible hint of lensing. Second, each expert inspected one by one all of the candidates selected in the previous step and grade them A,

† Available at `https://github.com/ClarkGuilty/CMUDeepLens-keras`
‡ Available at `https://github.com/ClarkGuilty/Qt-stamp-visualizer`

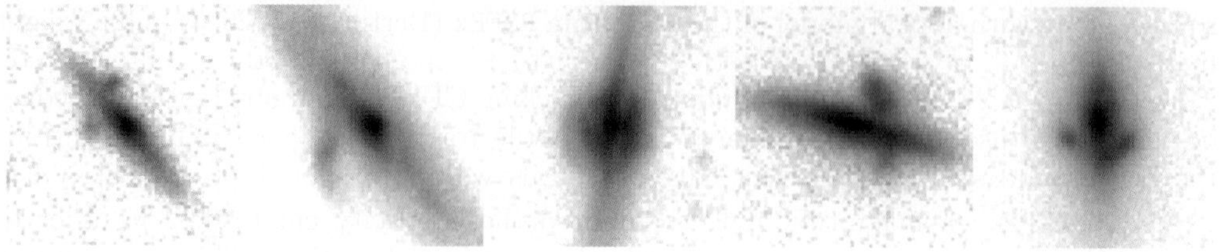

Figure 1. Examples of Grade A edge-on lens candidates recovered by the CNN. All of them are new candidates except for the one at the right (see J1559+3146 in Cañameras et al. 2020; Savary et al. 2022).

Table 1. Summary of the visual inspection of random stamps
Top: all types of lenses. Bottom: edge-on lenses only.

Magnitude bin	No. found	Estimated prevalence	Estimated No. of lenses (90% CI) [1]
17 to 18	3	1/10,000	33 [16, 82]
18 to 19	5	1/6,000	167 [88, 347]
19 to 20	3	1/10,000	285 [131, 734]
20 to 20.5	3	1/10,000	280 [128, 722]
17 to 18	1	1/30,000	11 [4, 50]
18 to 19	1	1/30,000	33 [12, 156]
19 to 20	0	< 1/13,100	< 218
20 to 20.5	0	< 1/13,100	< 214

Notes: [1] 90% confidence interval.

B, or X (A meaning sure lens; B, possibly a lens; and X, not a lens). Finally, all experts gathered together and decided a final list of candidates from all the objects graded A or B in the previous step.

4. Preliminary results and conclusion

Of the 6,978,977 original stamps, 52,088 had a CNN score larger than 0.5. Among those, we found 56 new lens candidates and rediscovered 5 previously reported ones. Of the new candidates, 40 (4 grade A and 36 grade B) had an edge-on disc galaxy as main deflector. The other 16 new candidates (3 grade A and 13 grade B) had LRGs or galaxy groups as main deflector. We present some examples of grade A edge-on lens candidates in Figure 1.

In order to correctly evaluate the performance of the CNN, we need to know the prevalence of *identifiable* edge-on lenses in the sample. To estimate it, we split the data set into magnitude bins and visually inspected 30,000 random objects per bin. Using Bayesian statistics, we estimated the prevalence of lenses of any type, and of edge-on lenses in particular, along with the number of identifiable candidates with a confidence interval of 90%. The results are summarized in Table 1.

Considering all types of lenses, we found candidates in every magnitude bin. The prevalence in all bins is consistent with 1 lens every 10,000 stamps. The estimated total number of lens candidates in the sample ranges from 363 to 1885, with 765 being the most likely value.

Edge-on lenses are found only in the brighter magnitude bins. For those bins, the estimated prevalence is 1 edge-on lens every 30,000 stamps. For the others, however, we can only provide an upper bound. Thus, the prevalence in the fainter bins is smaller than 1 edge-on lens for every 13,100 stamps. Finally, the estimated total number of edge-on lens candidates in the sample lies between 16 and 638.

Comparing the results of the CNN with the estimated prevalence of lenses in the sample, the number of edge-on candidates discovered by the CNN (40) is within the estimated range (16 to 638), albeit on the lower side. By contrast, the number of lens candidates of any type discovered by the CNN (56) is well outside of the estimated range (363 to 1885). This is to be expected since the CNN was trained to target edge-on lenses specifically, and they represent only a small fraction of all the lenses in the sample.

Finally, we combined the lens candidates discovered during the inspection of random stamps, with the candidates from the CNN, and the serendipitous discoveries, resulting in a total of 78 new lens candidates and 8 re-discoveries. Out of the new candidates, 48 (4 grade A and 44 grade B) had an edge-on disc galaxy as main deflector. The other 30 new candidates (7 grade A and 23 grade B) had LRGs, galaxy groups, or galaxy clusters as main deflector.

In conclusion, we showed the viability of finding edge-on lenses from single-band data in ground-based imaging surveys, in spite of their small Einstein radii. We discovered a sizable population of previously unknown edge-on lenses, along with some new candidates of lensing by LRG, galaxy groups, and galaxy clusters. Furthermore, we showcased a methodology to evaluate the performance of automated lens finders by estimating the prevalence of candidates in the studied sample, and comparing it to the number of candidates discovered by the lens finder. In this way we can evaluate the completeness of the lens search in an objective basis for that particular data set. Our work also illustrates the need for lens searches that explicitly target edge-on disc galaxies, and presents a methodology on how to carry them out. Our results and methods are particularly promising for the Euclid VIS instrument, which will have even sharper images and will observe 4 times the area studied in this work.

5. Acknowledgments

This work is supported by the Swiss National Science Foundation (SNSF). Based on observations obtained with MegaPrime/MegaCam, a joint project of CFHT and CEA/DAPNIA, at the Canada-France-Hawaii Telescope (CFHT) which is operated by the National Research Council (NRC) of Canada, the Institut National des Science de l'Univers of the Centre National de la Recherche Scientifique (CNRS) of France, and the University of Hawaii. The observations at the Canada-France-Hawaii Telescope were performed with care and respect from the summit of Maunakea which is a significant cultural and historic site.

References

A. Almeida et al. 2023, ApJS, 267(2), 44.

E. Bertin. In I. N. Evans, A. Accomazzi, D. J. Mink, and A. H. Rots, editors, *Astronomical Data Analysis Software and Systems XX* 2011, volume 442 of *Astronomical Society of the Pacific Conference Series*, 435.

E. Bertin 2013, Astrophysics Source Code Library, record ascl:1301.001.

E. Bertin and S. Arnouts 1996, A&AS, 117, 393–404.

E. Bertin and S. Arnouts 2010, Astrophysics Source Code Library, record ascl:1010.064.

S. Birrer and A. Amara 2018, *Physics of the Dark Universe*, 22, 189–201.

S. Birrer et al. 2021, *The Journal of Open Source Software*, 6(62), 3283.

B. J. Brewer et al. 2012, MNRAS, 422(4), 3574–3590.

R. Cañameras et al. 2021, A&A, 653, L6.

R. Cañameras et al. 2020, A&A, 644, A163.

F. Chollet and others 2018, Astrophysics Source Code Library, record ascl:1806.022.

A. A. Dutton et al. 2011, MNRAS, 417(3), 1621–1642.

I. Ferreras et al. 2010, MNRAS, 409(1), L30–L34.

S. D. J. Gwyn 2008, PASP, 120(864), 212.

F. Lanusse et al. 2018, MNRAS, 473(3), 3895–3906.

R. Li et al. 2021, ApJ, 923(1), 16.

J. F. Navarro, C. S. Frenk, and S. D. M. White 1996, ApJ, 462, 563.
K. Rojas et al. 2022, A&A, 668, A73.
E. Savary et al. 2022, A&A, 666, A1.
J. F. Sygnet, H. Tu, B. Fort, and R. Gavazzi 2010, A&A, 517, A25.
T. Treu et al. 2011, MNRAS, 417(3), 1601–1620.

Strong Gravitational Lensing in the Era of Big Data
Proceedings IAU Symposium No. 381, 2024
H. Stacey, A. Sonnenfeld & C. Grillo, eds.
doi:10.1017/S174392132300412X

Joint lensing–dynamics constraint on the elliptical galaxy mass profile from the largest galaxy–galaxy lens sample

Chin Yi Tan[1,2] and **Anowar Shajib**[3,2]

[1]Department of Physics, University of Chicago, Chicago, IL 60637, USA.
email: chinyi@uchicago.edu

[2]Kavli Institute for Cosmological Physics, University of Chicago, Chicago, IL 60637, USA

[3]Department of Astronomy & Astrophysics, The University of Chicago, Chicago, IL 60637, USA

Abstract. From various lensing and non-lensing observations, the total density profile in elliptical galaxies is well approximated by a power law mass distribution. However, as neither the dark matter nor the baryons individually follow the power law, this observational result has been referred to as the "bulge–halo conspiracy". We investigate the consistency of this conspiracy with higher precision than previous studies with the largest ever sample of galaxy–galaxy lenses assembled from archival *Hubble Space Telescope* (*HST*) data. By performing lens modelling with power law profiles and combining them with stellar dynamics using the mass-sheet transformation, we can constrain the profile's deviation from the power law model over a redshift range of 0.1 to 0.8. To uniformly model such a large sample of lens models, we use the automated modelling pipeline DOLPHIN. We also investigate the evolutionary trend of the total mass density slope of the lensing galaxies.

Keywords. gravitational lensing, galaxies: elliptical and lenticular, cD

1. Introduction

A galaxy-scale lens system is a lens configuration where a foreground (lensing) galaxy's mass causes the background (source) galaxy's light to be imaged into multiple lensed arcs or an Einstein ring. When the background galaxy of a galaxy-scale lens system does not contain a point source, we refer to the system as a galaxy–galaxy lens (as opposed to quasar lenses). As the lensed image is sensitive to the mass profile of the foreground galaxy, strong lensing at the galaxy scale can be used as a direct probe of the galaxies' mass distribution. For an overview of the many applications of galaxy–galaxy lenses in astrophysics and cosmology, see Shajib *et al.* (2022).

Both strong lensing and non-lensing observations of elliptical galaxies have found that a power law profile, $\rho(r) \propto r^{-\gamma}$, can be used as a good model for the total density profile of the galaxies, with the slope of the power law being observed to be slightly steeper than isothermal ($\gamma \sim 2.1$). This trend has been seen in observations using combinations of strong lensing, weak lensing, stellar dynamics, and X-ray observation (Treu *et al.* (2006), Cappellari *et al.* (2015), Humphrey *et al.* (2010), Gavazzi *et al.* (2007)). However, the individual dark matter and baryonic components of elliptical galaxies are not modelled with a power law, leading to the "bulge–halo conspiracy". Simulation of galaxy formulation has suggested that the power law-like nature of the total density profile can be explained through the redistribution of the mass distribution of the galaxy through collisionless accretion in gas-poor mergers (Remus *et al.* (2013)).

From N-body cosmological simulations, it is found that elliptical galaxies are formed from hierarchical mergers of dark matter haloes that also contain stars and gas. The types of merger processes experienced by the elliptical galaxies can be inferred through observations of the evolution of the total density profile of the galaxies. If the elliptical galaxies formed through dissipative processes via "wet" gas-rich mergers, the dark matter halo will contract, and we would expect the total density slope to steepen with decreasing redshift. However, suppose galaxies are formed through dissipation-less gas-poor mergers. In that case, we expect the opposite trend with the dark matter halo expanding and the total density slope shallowing with decreasing redshift.

Analyses based on strong lensing and stellar dynamics have measured a slightly steepening trend in the total density slope with decreasing redshift consistent with gas-rich mergers (Sonnenfeld *et al.* (2013), Bolton *et al.* (2012)). In comparison, cosmological simulations such as Illustris and Magneticum have found either a constant mass density slope or a slightly shallowing slope with decreasing redshift (Xu *et al.* (2017), Remus *et al.* (2017)). Remus *et al.* (2017) have found that this apparent disagreement disappears when we apply the same joint lensing and dynamics analysis to the simulated galaxies, which might indicate that systematic errors are present in the joint lensing and dynamics analysis.

Therefore, we assembled one of the largest uniformly modelled lens samples to investigate the above mentioned tensions. In Section 2, we first discuss the improvements made in our lens modelling pipeline compared to previous joint lensing and dynamics analysis. We then used our lens models to investigate the redshift trend of the total mass density slope of elliptical in Section 3. Finally, in Section 4, we combine our lens models with kinematic data to measure any deviation of our total density profiles with the power law.

2. Uniform Lens modelling

In many previous joint lensing and dynamics analyses, the lens image is first modelled with a singular isothermal ellipsoid (SIE) profile to obtain the Einstein radius. The mass density slope is then measured using kinematic data, assuming a power law model. However, using *HST*-quality imaging data, it is also possible to constrain the total density slope directly from images. Therefore, if we model *HST* images with the more flexible power-law ellipsoidal mass distribution (PEMD) model, we can extract all the information present in the lensed arcs. This also allows us to use the additional kinematic data to fit profiles beyond the power law.

We present Project Dinos, which aims to study the evolution in the properties of elliptical galaxies by remodelling archival *HST* images of the galaxy–galaxy lens systems with the more flexible PEMD model. Our lens samples comprised of a subset of lens systems from various surveys such as the Sloan Lens ACS (SLACS) survey, the Strong Lensing Legacy Survey (SL2S), and the BOSS Emission-Line Lens Survey (BELLS) based on the availability of visible-band *HST* imaging. To uniformly model such a large lens sample, we use the lens modelling pipeline DOLPHIN (Shajib *et al.* (2021)), which employs the lens modelling software LENSTRONOMY (Birrer *et al.* (2018), Birrer *et al.* (2021)) as its modelling engine. Whereas LENSTRONOMY contains many features to allow the user to fine-tune their lens models, DOLPHIN uses a semi-automated decision-making process and has a simplified user interface to reduce the time and effort needed for the user to model a lens system.

For the lensing galaxy, we set the mass profile to be the power-law ellipsoidal mass distribution (PEMD) while using two elliptical Sérsic profiles for the light profile. We then

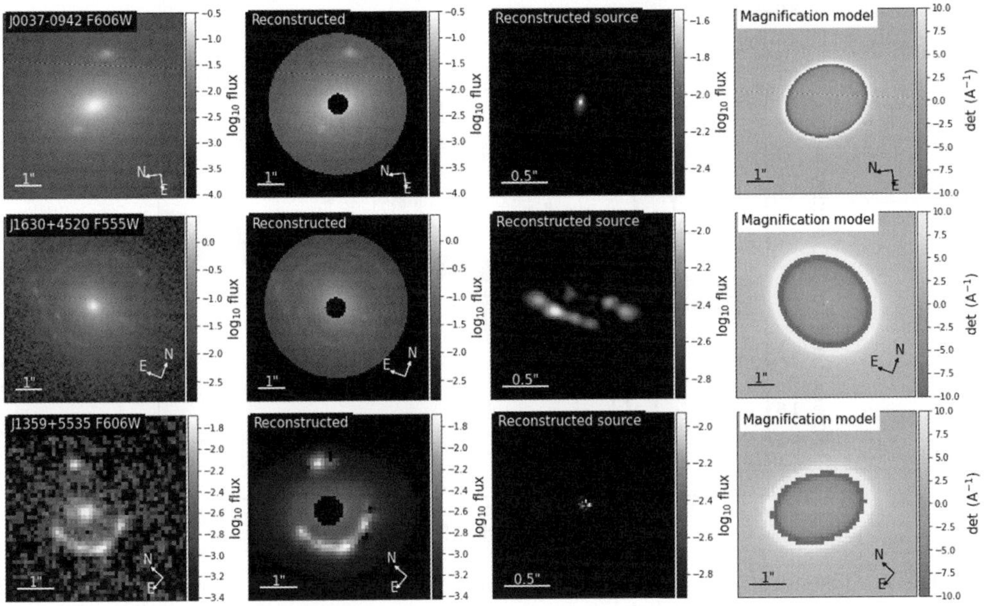

Figure 1. DOLPHIN lens models for selected SLACS (first two rows) and SL2S (last row) lenses. The first column shows *HST* images of the lenses in the visible band. The label in the first column indicates the name of the lens and band of the image. The second column shows the model's reconstructed image of the system, while the reconstructed source is shown in the third column. The lens model magnification map is shown in the fourth column.

use a combination of elliptical Sérsic profile and a basis set containing multiple shapelet components (Birrer *et al.* (2015)) to recreate the light profile of the source galaxy.

We use the Markov chain Monte Carlo (MCMC) method using EMCEE to obtain the posterior probability distribution function of the model parameters. Figure 1 shows examples of lens models from the SLACS and SL2S samples successfully modelled by DOLPHIN. A number of lens systems contain complicated lens features and they are not successfully reconstructed by the modelling pipeline. These lenses must be modelled on a case-by-case basis and thus are excluded from our analysis. Out of the initial 103 lens systems of SLACS, SL2S and BELLS lenses, we successfully model 77 systems.

3. The evolution of the total density slope

Before we combine our lens models with kinematic data, we first probe the evolution of the lensing galaxy's power-law mass density slope obtained from lensing-only information. Figure 2 shows the individual power-law mass density slope, $\gamma_{\rm PL}$, for each lens system plotted against the redshift their lensing galaxy. We fit a linear relation for the power-law mass density slope as a function of the redshift of the lensing galaxy, z, and the ratio of the effective radius of the galaxy's light profile to its Einstein radius, $R_{\rm eff}/\theta_{\rm E}$. In addition, we also allow for the linear fit to contain a redshift-independent intrinsic scatter σ_γ. The median power-law mass density slope for the combined population is measured to be

$$\langle \gamma_{\rm PL}(z) \rangle = (2.06 \pm 0.04) + (0.29 \pm 0.19)(z - 0.3) + (0.03 \pm 0.05)(\frac{R_{\rm eff}}{\theta_{\rm E}} - 1), \quad (1)$$

and the intrinsic scatter of power-law mass density slope is $\sigma_\gamma = 0.11 \pm 0.01$. Therefore, we found that the total mass density slope shallows with decreasing redshift at a significance of 1.5σ. This is opposite to the trend with previous joint lensing and dynamics analyses (Sonnenfeld *et al.* (2013), Bolton *et al.* (2012)) but consistent with comological simulation.

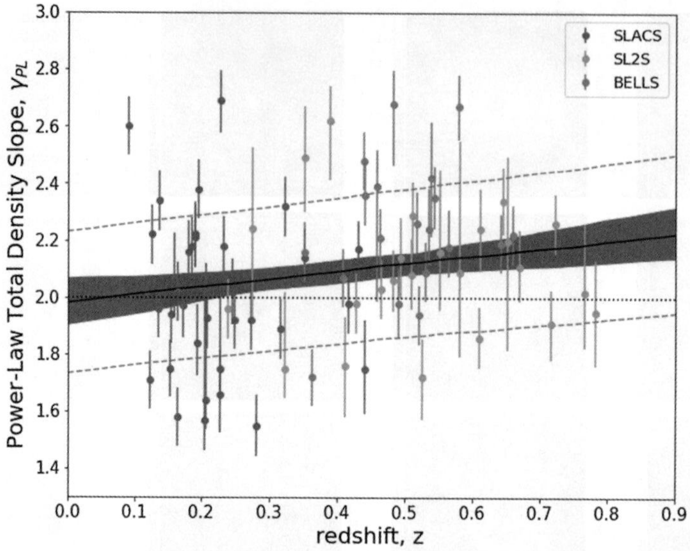

Figure 2. The evolution of the power-law mass density slope, $\gamma_{\rm PL}$, over the redshift of the lensing galaxy, z. The coloured points represent the measurement and uncertainty of $\gamma_{\rm PL}$ for each individual lens system obtained from DOLPHIN. The whole black line traces the median fit for $\langle\gamma_{\rm PL}\rangle$, while the grey shaded regions represent the 1σ credible region both marginalised over the $R_{\rm eff}/\theta_{\rm E}$ dependence. The intrinsic scatter, σ_γ, is represented by the dashed lines.

However, due to the large intrinsic scatter found in the fits, a larger lens model sample is needed to increase the statistical power of the analysis.

4. Measuring the deviation from power-law with the mass-sheet transform

The mass-sheet transform (MST) (Falco *et al.* (1985)) is a transformation that changes the shape of the lensing galaxy's profile while preserving all its lensing observables. Under the MST, the convergence field of the lensing galaxy, κ, transforms as,

$$\kappa \rightarrow \kappa' = \lambda_{\rm int}\kappa + (1 - \lambda_{\rm int}), \tag{2}$$

where $\lambda_{\rm int}$ is the internal MST parameter for the transformation. For velocity dispersion measurements obtained from a single aperture, the velocity dispersion approximately scales with the internal MST parameter (Birrer *et al.* (2020)) as

$$\sigma \rightarrow \sqrt{\lambda_{\rm int}}\sigma. \tag{3}$$

Therefore, by comparing velocity dispersion predicted by our power-law-based lens models to the measured dispersion from spectroscopy, we can measure the MST parameter, $\lambda_{\rm int}$, and thus probe the deviation from the power law.

We use the hierarchical strong lensing systems analysis pipeline HIERARC (Birrer *et al.* (2020)) to provide a population-level estimate of $\lambda_{\rm int}$. We also also incorporate the external convergence measurements, $\kappa_{\rm ext}$, from the line-of-sight galaxies and stellar anisotropy of the lensing galaxy using the Osipkov-Merritt formalism in our analysis. Similar to the evolution of the power-law mass density slope, we fit a linear relation for $\lambda_{\rm int}$ as a function of the redshift of the lensing galaxy, z, and the ratio of the effective radius of the galaxy's light profile to its Einstein radius, $R_{\rm eff}/\theta_{\rm E}$, in the form of

$$\langle\lambda_{\rm int}\rangle = \lambda_{\rm int}^{\rm ref} + \alpha_\lambda(z - 0.3) + \beta_\lambda(\frac{R_{\rm eff}}{\theta_{\rm E}} - 1). \tag{4}$$

Tan *et al.* (in prep) will present the values of the associated parameters and provide discussion on its implication for the evidence of the deviation of the total mass profile from the power law. In the near future, we intend to extend our analysis to a composite model consisting of both dark matter and baryonic components using techniques similar to those employed by Shajib *et al.* (2021) to obtain a more physically motivated mass profile of elliptical galaxies beyond the power law.

5. Acknowledgements

This project and a portion CYT's travel to the symposium were supported by NASA through the NASA Hubble Fellowship grant HST-HF2-51492 awarded to AJS by the Space Telescope Science Institute, which is operated by the Association of Universities for Research in Astronomy, Inc., for NASA, under contract NAS5-26555. Another portion of CYT's travel cost was supported by the Department of Energy grant DE-AC02-07CH11359 subcontract 6749003 at Fermilab and the University of Chicago.

References

Birrer S., Amara A., Refregier A. 2015, *ApJ*, 813, 102
Birrer S., Amara A. 2018, *Physics of the Dark Universe*, 22, 189
Birrer S., et al. 2020, *A&A*, 643, A165
Birrer S., et al. 2020, *JOSS*, 6, 3283
Bolton A., et al. 2012, *ApJ*, 757, 82
Cappellari M., et al. 2015, *ApJ*, 804, L21
Falco E. E., Gorenstein M. V., Shapiro I. I. 2015, *ApJ*, 289, L1
Gavazzi R., et al. 2007, *ApJ*, 667, 176
Humphrey P. J., Buote D. A. 2010, *MNRAS*, 403, 2143
Remus R.-S., et al. 2013, *ApJ*, 766(2), 71
Remus R.-S., et al. 2017, *MNRAS*, 464(3), 3742–3756
Shajib A. J., et al. 2022, arXiv e-prints, p. arXiv:2210.10790
Shajib A.J., Treu T., Birrer S., Sonnenfeld A., 2021, *MNRAS*, 503(2), 2380–2405
Sonnenfeld A., et al. 2013, *ApJ*, 777(2), 98
Treu T., Koopmans L. V., Bolton A. S., Burles S., Moustakas L. A., 2006, *ApJ*, 640, 662
Xu, D., *et al.* 2017, *MNRAS*, 469(2), 1824–1848

Strong Gravitational Lensing in the Era of Big Data
Proceedings IAU Symposium No. 381, 2024
H. Stacey, A. Sonnenfeld & C. Grillo, eds.
doi:10.1017/S1743921323003897

Strong Lens Detection 2.0: Machine Learning and Transformer Models

Hareesh Thuruthipilly

National Centre for Nuclear Research, Astrophysics division
ul. Pasteura 7, 02-093 Warszawa
email: hareesh.thuruthipilly@ncbj.gov.pl

Abstract. Upcoming large-scale surveys like LSST are expected to uncover approximately 10^5 strong gravitational lenses within massive datasets. Traditional manual techniques are too time-consuming and impractical for such volumes of data. Consequently, machine learning methods have emerged as an alternative. In our prior work (Thuruthipilly et al. 2022), we introduced a self-attention-based machine learning model (transformers) for detecting strong gravitational lenses in simulated data from the Bologna Lens Challenge. These models offer advantages over simpler convolutional neural networks (CNNs) and competitive performance compared to state-of-the-art CNN models. We applied this model to the datasets from Bologna Lens Challenge 1 and 2 and simulated data on Euclid.

Keywords. Gravitational strong lensing, data analysis, image processing, observations

1. Introduction

Strong gravitational lensing is a fascinating prediction of General Relativity (GR). It occurs when a foreground galaxy's massive gravity bends light from a distant source, forming arcs or multiple images. Since the first galaxy-galaxy strong lensing system discovery by Hewitt et al. (1988), numerous applications in cosmology and astrophysics have emerged for strong lenses (SLs). However, strong gravitational lensing is rare, with roughly one lens per thousand galaxies observed. Advanced missions like Euclid (R. et al. 2021) and Legacy Survey of Space and Time (LSST) (Ivezić et al. 2019) are expected to identify about 10^5 strong lensing events among 10^9 galaxies. Analyzing data manually or semi-automatically in future large-scale surveys is impractical and time-consuming, necessitating faster, reliable, and robust alternative approaches.

Recent advancements in natural language processing (NLP) led to the development of transformer models, a breakthrough in NLP introduced by Vaswani (2017). While these transformers have revolutionized machine learning, their application in astrophysics remains largely unexplored. In this study, we investigate the potential of attention-based models, specifically transformers, to identify strong gravitational lenses (SLs). We implemented transformers on Bologna Lens Challenge 1 and 2 data and also assessed their performance on simulated data for the Euclid mission. Additionally, we compared the performance of transformer models with other top-performing models on the same dataset.

2. Methods

We base our Strong Lens (SL) detection Transformer models on self-attention principles. Self-attention, generally expressed as $\text{Attention}(Q, K, V) = \text{softmax}\left(QK^T/\sqrt{d_k}\right) V$, assigns importance to input features based on their inherent relationships, where Q, K, V

Table 1. 'Lens Detector 15' vs. Best CNNs in the Literature.

Dataset	Model	AUROC
Bologna Lens Challenge 1	Lens Detector 15	0.98
	CMU-DeepLens (Metcalf et al. 2019)	0.98
Bologna Lens Challenge 2	Lens Detector 15	**0.83**
	EfficientNets (Bom et al. 2022)	0.82
Euclid Simulation Data S1	Lens Detector 15	**0.98**
	Inception (Euclid Collaboration 2023)	0.92
Euclid Simulation Data S2	Lens Detector 15	**0.98**
	Inception (Euclid Collaboration 2023)	0.88
Euclid Simulation Data S3	Lens Detector 15	**0.98**
	Inception (Euclid Collaboration 2023)	0.90
Euclid Simulation Data S4	Lens Detector 15	**0.97**
	Inception (Euclid Collaboration 2023)	0.81

are vectors and $\sqrt{d_k}$ represents the key vector's dimension. In our study, we consistently employ the 'Lens Detector 15' model mentioned in Thuruthipilly et al. (2022) across three different datasets: Bologna Lens Challenge 1, 2, and the Euclid simulation dataset. For the Bologna Lens Challenge 1 dataset, we trained this model with randomly initialized weights. However, we utilized pre-trained weights from our training on the Bologna Lens Challenge 1 for the other two datasets.

3. Results and Conclusions

In Table 1, we conduct a comparative evaluation of the 'Lens Detector 15' transformer model's performance in contrast to the best-performing convolutional neural networks (CNNs) documented in existing literature across diverse simulated datasets. Notably, the transformer models demonstrate exceptional performance compared to the CNNs mentioned in prior studies, highlighting the adaptability of transformers in the context of strong lens detection. For a comprehensive understanding of the datasets, please refer to Metcalf et al. (2019); Bom et al. (2022) for Bologna Lens Challenge 1 and 2, and Euclid Collaboration (2023) for the Euclid simulation datasets.

References

Bom et al. 2022, Developing a victorious strategy to the second strong gravitational lensing data challenge. *MNRAS*, 515(4), 5121–5134.

Euclid Collaboration 2023, Euclid Preparation TBD. Characterization of convolutional neural networks for the identification of galaxy-galaxy strong lensing events. *arXiv e-prints*, arXiv:2307.08736.

Hewitt, J. N., Turner, E. L., Schneider, D. P., Burke, B. F., & Langston, G. I. 1988, Unusual radio source MG1131+0456: a possible Einstein ring. *Natur*, 333(6173), 537–540.

Ivezić, Ž., Kahn, S. M., Tyson, J. A., Abel, B., Acosta, E., Allsman, R., Alonso, D., AlSayyad, Y., Anderson, S. F., Andrew, J., & et al. 2019, Lsst: From science drivers to reference design and anticipated data products. *ApJ*, 873(2), 111.

Metcalf, R. B., Meneghetti, M., Avestruz, C., Bellagamba, F., Bom, C. R., Bertin, E., Cabanac, R., Courbin, F., Davies, A., Decencière, E., & et al. 2019, The strong gravitational lens finding challenge. *Astronomy & Astrophysics*, 625, A119.

R., S., J., A., Y., M., C., B., & S., C. C. 2021, Euclid preparation: I. The Euclid Wide Survey. *arXiv e-prints*, arXiv:2108.01201.

Thuruthipilly, H., Zadrozny, A., Pollo, A., & Biesiada, M. 2022, Finding strong gravitational lenses through self-attention. Study based on the Bologna Lens Challenge. *Astronomy & Astrophysics*, 664, A4.

Vaswani, A. Attention is all you need. In *Advances in Neural Information Processing Systems 30: Annual Conference on Neural Information Processing Systems 2017, December 4-9, 2017, Long Beach, CA, USA* 2017, pp. 5998–6008.

Strong Gravitational Lensing in the Era of Big Data
Proceedings IAU Symposium No. 381, 2024
H. Stacey, A. Sonnenfeld & C. Grillo, eds.
doi:10.1017/S1743921323004222

The last stand before Rubin: semi-automated inverse modelling of galaxy-galaxy strong lensing systems

João Paulo C. França[1,†], Martin Makler[2,1], Ingrid Beloto[3], Eduardo Cypriano[3], Renan A. Oliveira[4], Thiago S. Gonçalves[5] and James Nightingale[6,7]

[1]Centro Brasileiro de Pesquisas Físicas

[2]International Center for Advanced Studies & Instituto de Ciencias Físicas, ECyT-UNSAM & CONICET

[3]Instituto de Astronomia, Geofísica e Ciências Atmosféricas, Universidade de São Paulo

[4]Centro de Ciências Exatas, Universidade Federal do Esprito Santo

[5]Observatório do Valongo, Universidade Federal do Rio de Janeiro

[6]Department of Physics, Centre for Extragalactic Astronomy, Durham University

[7]Department of Physics, Institute for Computational Cosmology, Durham University

Abstract. Galaxy-galaxy strong lensing (SL) systems provide a unique opportunity to test modified gravity theories. Deviations from General Relativity are encoded in the post-Newtonian parameter (γ). As a preparation for the upcoming data from the Vera Rubin Observatory Legacy Survey of Space and Time (LSST), our research group collected imaging data of SL systems from ground-based telescopes and conducted spectroscopic observations of 21 systems on the Southern Astrophysical Research (SOAR) Telescope to measure the lens velocity dispersions, σ_v. We briefly describe the semi-automated SL modelling of the systems in this sample and combine the results with σ_v from SOAR to derive an estimate for γ. Our preliminary results yield a value of $\gamma = 1.17^{+0.29}_{-0.33}$, which is consistent with General Relativity. Although the error bars are limited by the sample size, this result represents the first constraint on modified gravity obtained purely from ground-based data, with a sample completely independent from previous studies, and which allows for a self consistent end-to-end analysis.

Keywords. gravitational lensing: strong - methods: data analysis - galaxies: structure - techniques: spectroscopic

---

1. SOAR sample for measuring the γ post-Newtonian parameter

Previous studies have used galaxy-galaxy strongly lensed systems (aka "Einstein rings") to set constraints on the post-Newtonian parameter γ, which is essentially the ratio between the two scalar potentials that appear in the space-time metric at the linear level (the so-called Newtonian and curvature potentials). This ratio can be constrained by comparing the trajectories of massless bodies (photons), probed by gravitational lensing, to the motion of non-relativistic particles, probed by the stellar dynamics. The first is sensitive to the sum of the potentials, while the second is sensitive only to the Newtonian

† joaofranca@cbpf.br

potential. Therefore, by comparing lensing mass reconstructions to the stellar velocity dispersion in galaxies one is able to set constraints on γ (see Schwab, Bolton, and Rappaport (2009), Cao et al. (2016), Liu et al. (2022)). While the most recent studies contain over 150 systems, all previous estimates of γ build up over the previous samples (i.e. they all share a common set of systems). Furthermore the vast majority of the systems are modelled from space-based (i.g. HST) data, from surveys such as SLACS (Auger et al. 2010) and BELLS (Bolton et al. 2012) (see also Shajib et al. (2021), Etherington et al. (2021)). Typically, the systems from these samples have velocity dispersions from Sloan Digital Survey (SDSS) spectroscopic data.

In this work, we are interested in using systems that are modelled directly from ground-based images, as expected in the future from LSST. To this end our group has carried out an extensive compilation of systems in the footprint to the current ground-based wide-field imaging surveys that achieve typical resolutions around 1 arcsecond or better. This sample includes, but is not limited to, imaging data from the Survey of Gravitationally-lensed Objects in HSC Imaging (SuGOHI), Dark Energy Survey (DES), Kilo-Degree Survey (KiDS), and the Legacy Survey. This sample contains over 30,000 candidate gravitational lens systems, among which a few hundred are expected to be suitable for modelling. A detailed description of this dataset will be presented in Oliveira et al., in prep. We refer to this comprehensive compilation of gravitational lenses as "The Last Stand Before Rubin" (LaStBeRu).

With the aim of inferring the post-Newtonian parameter, we focused on obtaining good estimates of the line-of-sight velocity dispersion σ_v for Einstein rings selected from the LaStBeRu sample, thus building a new dataset for measuring γ. We selected systems which had source and lens redshifts available from the literature and whose lens galaxies were bright enough to yield spectra with† $S/N \sim 10 - 20$ (in a 4-m class telescope), to allow for a good determinations of σ_v. We visually inspected all systems that passed our cut to select those we expected to enable a good strong lensing modelling from the ground based data. Finally, we required good observability conditions from the south. Our selected sample was observed with the Goodman high throughput spectrograph in single slit mode on the SOAR telescope. Our sample contains 21 systems with σ_v determined using the pPXF method on the reduced SOAR spectra.

2. Modelling results and modified gravity constraints

To model the imaging data from the 21 lensing systems, we initiated the analysis by deriving noise maps and point-spread functions (PSFs). Noise maps are estimated from the background levels of the images. With the recovered background values, we scaled the original image according to the inverse of the effective gain, which is determined by the product of gain, the number of combined images, and exposure time. For the PSFs, we identified the astronomical objects in the wide-field images using the SExtractor software, selecting star candidates by applying a cutoff of 0.8 on the stellarity parameter. From the resulting objects, we selected the stars closest to the gravitational lens and applied the interactive PSFr reduction pipeline (Birrer et al., in prep.), which utilizes features from lenstronomy (Birrer et al. 2021).

The lensing systems are modelled using an updated version of the Source Lens And Mass (SLAM) pipeline (which builds up upon Etherington et al. (2021)), where, by default, the lens light brightness is fitted concurrently with the lens mass and source brightness. The method begins by parametrically fitting the source to provide an initial estimate of the modelling. Next, we fits the source using a non-uniform distribution of pixels and re-fit the lens brightness. We finalize the pipeline by fitting a

† Ideally, we want to achieve a higher signal-to-noise ratio than SDSS (Sloan Digital Sky Survey) because σ_v dominates the balance of statistical error in γ.

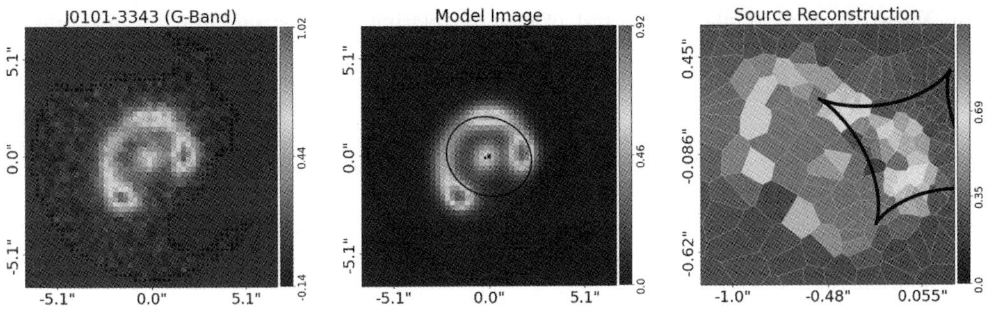

Figure 1. Modelling results for system J0101-3343. Left: g-band imaging data from DES used for the modelling. Middle: reconstructed image. The black curve shows the critical line. Right: zoomed-in reconstruction of the source plane. The black curve shows the caustic.

power-law mass profile. These steps were implemented using features from PyAutoLens (Nightingale, Dye, and Massey 2018; Nightingale et al. 2021). If the light distribution of the system is complex, for instance, due to the presence of nearby galaxies or blending of the arc with other sources, we employ an interactive extraction method that utilizes segmentation maps generated via SExtractor to create masks for removing the unwanted sources. We show in Fig. 1 an example of the modelling results for object J0101-3343.

Combining the results from our modelling on the 21 systems with the velocity dispersions obtained from the SOAR data and following the same methodology as in Cao et al. (2016) we were able to derive a first estimate of the post-Newtonian parameter: $\gamma = 1.17^{+0.29}_{-0.33}$. In General Relativity $\gamma = 1$, which is consistent with our results at the 68% confidence level. Although the error bars are significantly larger than the current estimates in the literature, to the best of our knowledge, this is the first result using a sample of systems that is completely independent from the one used in previous works. More importantly, these results show the potential for using ground based data to test gravity. Furthermore, the current data allows for a self consistent end-to-end analysis of the data, without the need to set strong global priors on the mass and light distribution of the lenses, as is usually done, and enabling us to consider the effect of the PSF on a system-by-system basis. Such analysis is expected to be less prone to systematics and will be explored in forthcoming studies.

Supplementary material

To view supplementary material for this article, please visit http://dx.doi.org/10.1017/S1743921323004222.

References

Auger, M. W., et al., The Sloan Lens ACS Survey. X. Stellar, dynamical, and total mass correlations of massive early-type galaxies. The Astrophysical Journal 724.1 (2010): 511.

Birrer, Simon, et al., lenstronomy II: A gravitational lensing software ecosystem. Journal of Open Source Software, 6(62), 3283.

Bolton, Adam S., et al., The BOSS emission-line lens survey. II. Investigating mass-density profile evolution in the SLACS+ BELLS strong gravitational lens sample. The Astrophysical Journal 757.1 (2012): 82.

Cao, Shuo, et al., Limits on the power-law mass and luminosity density profiles of elliptical galaxies from gravitational lensing systems. Monthly Notices of the Royal Astronomical Society 461.2 (2016): 2192–2199.

Etherington, Amy, et al., Automated galaxy-galaxy strong lens modelling: No lens left behind. Monthly notices of the Royal Astronomical Society 517.3 (2022): 3275–3302.

Liu, Xiao-Hui, et al., Galaxy-scale test of general relativity with strong gravitational lensing. The Astrophysical Journal 927.1 (2022): 28.

Nightingale, J. W., Simon Dye, and Richard J. Massey, AutoLens: automated modeling of a strong lens's light, mass, and source. Monthly Notices of the Royal Astronomical Society 478.4 (2018): 4738–4784.

Nightingale, J. W., et al., PyAutoLens: Open-Source Strong Gravitational Lensing. Journal of Open Source Software, 6(58), 2825.

Schwab, Josiah, Adam S. Bolton, and Saul A. Rappaport, Galaxy-scale strong-lensing tests of gravity and geometric cosmology: constraints and systematic limitations. The Astrophysical Journal 708.1 (2009): 750.

Shajib, Anowar J., et al., Dark matter haloes of massive elliptical galaxies at $z \sim 0.2$ are well described by the Navarro-Frenk-White profile. Monthly Notices of the Royal Astronomical Society 503.2 (2021): 2380–2405.

Strong Gravitational Lensing in the Era of Big Data
Proceedings IAU Symposium No. 381, 2024
H. Stacey, A. Sonnenfeld & C. Grillo, eds.
doi:10.1017/S1743921323003708

Lens Discovery in the Era of Wide-area Surveys

Philip Holloway[1], **Philip J Marshall**[2,3] **and Aprajita Verma**[1]

[1]Sub-department of Astrophysics, University of Oxford, Denys Wilkinson Building, Keble Road, Oxford, OX1 3RH, UK. email: philip.holloway@physics.ox.ac.uk

[2]Kavli Institute for Particle Astrophysics and Cosmology, Department of Physics, Stanford University, Stanford, CA 94305, USA

[3]SLAC National Accelerator Laboratory, Menlo Park, CA 94025, USA

Abstract. Forthcoming data from the Vera Rubin Observatory, *Euclid* and *Roman* telescopes are expected to increase the number of strong lenses by two orders of magnitude. With current discovery methods these would be accompanied by an even greater number of false positives. In that context we find that using an ensemble of classifiers would provide a more complete sample of high-purity lenses and present methods to post-process the outputs of such classifiers to give reliable probabilities that a given image contains a lens.

Keywords. gravitational lensing: strong, galaxies: general, methods: data analysis

1. Introduction

With the arrival of wide-area telescopes such as the Vera Rubin Observatory, Euclid and Roman space telescopes, the number of strong lenses identified will increase to $\sim 10^5$ (Collett (2015), Holloway *et al.* (2023)). Current lens detection techniques require significant time-investment to remove false positives identified by automated or human classifiers which will only increase when such telescopes come online. This motivates our investigation into methods for improving the performance of strong lens finding.

2. Data and Method

To develop and test our method, we use outputs from two strong lens classifiers applied to Hyper-Suprime Cam (HSC) data. These classifiers are 1) a neural network (HOLISMOKES VI, Cañameras *et al.* (2021)) and 2) citizen science classifications from Space Warps (SuGOHI VI, Sonnenfeld *et al.* (2020)). These were cross-matched to within 1", producing $\sim 110,000$ galaxies for which outputs from both classifiers were available. For a subset of 3,514 typically high-scoring objects, grades were available following subsequent visual inspection. We used these grades as a 'ground-truth' to determine the performance of each classifier.

We mapped the classifier output to calibrated probability as follows. We took the distribution of grade A+B candidates (here considered true lenses) as a function of classifier output ranking and applied the following procedures to determine this mapping for each classifier: isotonic regression (Zadrozny & Elkan (2002)), variable bin fitting (akin to a moving average, but with a fixed number of lenses per bin), and the Kullback-Leibler Importance Estimation Procedure (KLIEP, Sugiyama *et al.* (2008)). These calibration mappings are validated against a separate validation set in Figures 1a & 1b which show

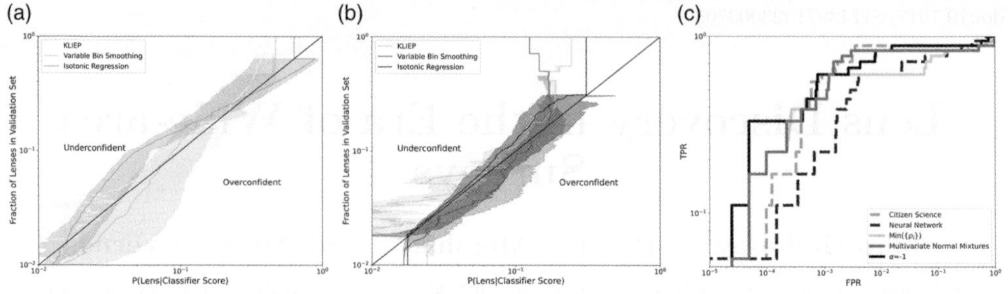

Figure 1. a & b) Validation for the 3 calibration methods applied to (a) the citizen science classifier and (b) the neural network: a perfectly calibrated classifier would lie along the $y = x$ line. c) Receiver operating characteristic (ROC) curve for the individual lens classifiers (dashed) and combined methods. Ungraded subjects were treated as non-lenses for this plot.

that across a wide region of probability-space, such a mapping can provide an accurate probability a given object is a lens.

To combine the calibrated outputs we used a generalised mean of the form: $P(p_i, \alpha) = (\frac{1}{N}\sum_i^N p_i^\alpha)^{\frac{1}{\alpha}}$ where p_i denotes the calibrated outputs from the N different classifiers, and α is a tuneable parameter; $\alpha \to \pm\infty$ corresponds to $Max(\{p_i, ...p_N\})$ and $Min(\{p_i, ...p_N\})$. We also trialled Bayesian probability combination using multivariate normal mixtures detailed in Pirš & Štrumbelj (2019). The results of the best performing methods are shown in Figure 1c. We find at the high-purity end of the curve (FPR$\sim 10^{-4}$), combining the two classifiers can roughly double the expected completeness from $\sim 15\%$ to $\sim 30\%$ though the ensemble classifier doesn't improve significantly that of the citizen science for larger false positive rates. With forthcoming wide-area surveys, this will help reduce the amount of time-consuming visual inspection required to verify lens candidates. A more in-depth analysis and further results will be presented in a forthcoming paper, Holloway *et al.* (2023, in prep).

3. Conclusion

We find the outputs of strong lens classifiers can be post-processed to give reliable probabilities that a given image contains a lens allowing the accurate ranking of lens candidates identified across multiple different methods. We find an ensemble classifier would provide a more complete sample of lenses for a given (high) value of purity than the individual classifiers by themselves. This will allow a larger sample of strong lenses to be used for future population-level analysis enabled by LSST, *Euclid* and *Roman* surveys.

4. Acknowledgements

With thanks to Raoul Cañameras and Alessandro Sonnenfeld for providing data for this project.

Supplementary material

To view supplementary material for this article, please visit http://dx.doi.org/10.1017/S1743921323003708

References

Collett, T. E. 2015, *ApJ*, 811, 20
Holloway, P., Verma, A., Marshall, P. J., More, A., Tecza, M. 2023, *MNRAS*, 525, 2341–2354
Cañameras, R., Schuldt, S., Shu, Y., *et al.* 2021 *A&A*, 653, L6

Sonnenfeld, A., Verma, A., More, A., *et al.* 2020 *A&A*, 642, A148
Zadrozny, B., Elkan, C. 2002, Proceedings of the ACM SIGKDD International Conference on Knowledge Discovery and Data Mining, 694–699
Sugiyama, M., Suzuki, T., Nakajima, S. *et al.* 2008, Annals of the Institute of Statistical Mathematics, 60, 699
Pirš, G., Štrumbelj, E. 2019, Journal of Machine Learning Research, 20, 1892–1909

Chapter 2

Strong Gravitational Lensing in the Era of Big Data
Proceedings IAU Symposium No. 381, 2024
H. Stacey, A. Sonnenfeld & C. Grillo, eds.
doi:10.1017/S1743921323004027

Gravitational lenses in hydrodynamical simulations

Giulia Despali[1,2], **Felix M. Heinze**[3] **and Claudio Mastromarino**[4,5]

[1]Alma Mater Studiorum - Universitá di Bologna, Dipartimento di Fisica e Astronomia "Augusto Righi", Via Gobetti 93/2, Bologna, Italy

[2]INAF - Osservatorio di Astrofisica e Scienza dello Spazio di Bologna, via Gobetti 93/3, I-40129, Bologna, Italy

[3]Universität Heidelberg, Zentrum für Astronomie, Institut für Theoretische Astrophysik, Albert-Ueberle-Straße 2, D-69120 Heidelberg, Germany

[4]Universitá degli studi di Roma 'Tor Vergata', Via della Ricerca Scientifica, 1, 00133, Roma, Italy

[5]INFN-Sezione di Roma 'Tor Vergata', Via della Ricerca Scientifica, 1, 00133, Roma, Italy

email: `giulia.despali@unibo.it`

Abstract. The gravitational lensing signal produced by a galaxy or a galaxy cluster is determined by its total matter distribution, providing us with a way to directly constrain their dark matter content. State-of-the-art numerical simulations successfully reproduce many observed properties of galaxies and can be used as a source of mock observations and predictions. Many gravitational lensing studies aim at constraining the nature of dark matter, discriminating between cold dark matter and alternative models. However, many past results are based on the comparison to simulations that did not include baryonic physics. Here we show that the presence of baryons can significantly alter the predictions: we look at the structural properties (profiles and shapes) of elliptical galaxies and at the inner density slope of subhaloes. Our results demonstrate that future simulations must model the interplay between baryons and alternative dark matter, to generate realistic predictions that could significantly modify the current constraints.

Keywords. strong lensing, cosmology, dark matter, numerical simulations

1. Introduction

One of the foundations of the concordance cosmological model is that approximately 85 percent of the universe's matter content is in the form of some yet unknown component that we can detect only through its gravitational effect: dark matter. Starting from small density perturbations in the early universe, dark matter leads the gravitational collapse on large scales, forming the cosmic-web structure of filaments along which gas flows to the potential wells of dark matter haloes, where it cools and creates stars - eventually leading to the formation of galaxies. Despite decades of experimental searches, there is yet to be a definitive answer to the most important and long-standing question in cosmology: the nature of dark matter. The favored particle candidates for cold dark matter (CDM) - e.g. axions and WIMPS, weakly interacting massive particles - have so far not been detected, motivating the study of alternative models that are allowed by particle physics assumptions and span a wide range of masses - from the eV to the GeV scale. Strong gravitational lensing is one of the most promising techniques to study the dark matter

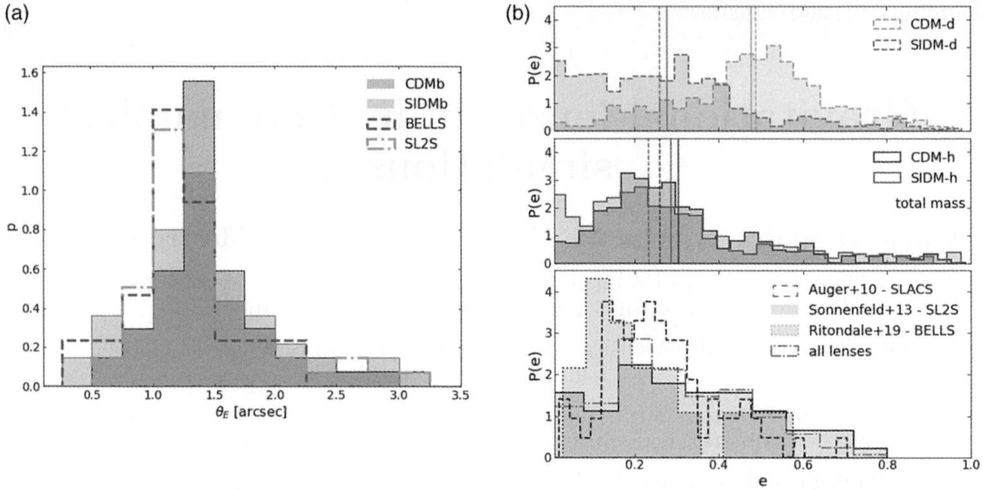

Figure 1. Properties of gravitational lenses in hydrodynamical simulations of CDM and SIDM universes. (*a*) Distribution of the Einstein radii θ_E of elliptical galaxies, measured in cold (gray) and self-interacting (red) dark matter simulations, compared to those of observed galaxy lenses from the BELLS (blue line) and SL2S (orange line) surveys. (*b*) Distribution of projected ellipticities e measured at $r = 10$ kpc from the halo centre. The top and middle panel shows the results from dark-matter-only and hydrodynamical simulations, while the bottom panel reports observational measurements from gravitational lensing data sets. The figures are adapted from Mastromarino et al. (2023) and Despali et al. (2022).

distribution of galaxies, given its ability to measure the total density distribution (both luminous and dark) in their central regions.

In the innermost regions of haloes and subhaloes, both self-interacting dark matter (SIDM) and baryons can alter the density profile in these regions. State-of-the-art hydrodynamical simulations can successfully reproduce galaxy formation and are thus an essential tool in the comparison and interpretation of lensing observations. However, constraints on alternative dark matter models have often been set by comparing observations to simulated predictions that do not include the physics of baryons. Recently, new numerical efforts have proved that it is essential to include the physics of baryons in simulations and correctly take it into account in comparison with observation. Here we present two such cases, relevant for gravitational lensing studies.

2. Overview

2.1. *Lens galaxies in SIDM universes*

Massive systems such as elliptical galaxies and galaxy groups are the typical strong lenses at z 0.2-1: Peter et al. (2013) derived stringent constraints ($\sigma_T/m_\chi \simeq 0.1cm^2g^{-1}$) on the self-interaction cross-section by comparing the lensing and X-ray signal in observed systems of this mass range to dark-matter-only simulations. Our recent works based on small samples of galaxies have shown that including treatment for baryonic physics can alleviate these constraints. Mastromarino et al. (2023) analysed the Eagle50-SIDM simulation by Robertson et al. (2021) and compared the density profiles and lensing signal of massive galaxies found in the CDM and SIDM runs. These are cosmological boxes (50 Mpc on a side) that have been re-simulated in two SIDM scenarios: (*i*) a model with constant cross-section $\sigma = 1cm^2g^{-1}$ and (*ii*) a model where the cross-section is velocity dependent. While all dark matter haloes ($10^{12}M_\odot \leq M \leq 10^{14}M_\odot$) form a central core in the dark-matter-only runs, the cuspy baryonic profile of the central galaxy in

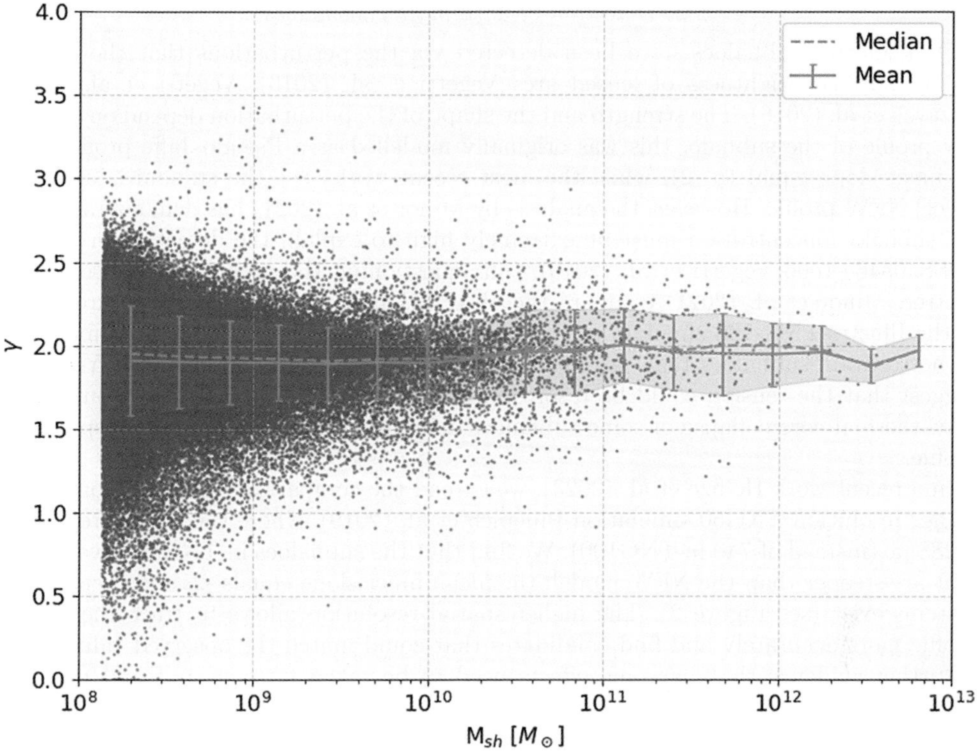

Figure 2. Slope γ of the total density profiles of subhaloes in the TNG50-1 hydrodynamical simulation, as a function of subhalo mass. The mean and median values for each mass bin are represented by the solid and dashed orange lines. The Figure is adapted from Heinze et al. (2023).

the hydrodynamical runs prevents the core formation and the CDM and SIDM profiles differ by less than 20 percent. Correspondingly, the Einstein radii calculated from the simulated galaxies follow the distribution shown in the left panel of Figure 1. Both models successfully match the observed Einstein radii from the SLACS Bolton et al. (2008) and SL2S surveys Sonnenfeld et al. (2013). This demonstrates that a cross-section of $\sigma = 1cm^2g^{-1}$ is still compatible with lensing observations: a small effect of self-interactions at this mass scale can be expected given that the baryonic contribution to the central density is maximised for galaxies with $M \simeq 10^{13}M_\odot$.

Similarly, the inclusion of baryons also modifies the distribution of halo shapes. In dark-matter-only simulations, CDM haloes are on average prolate and more elongated towards the centre, while self-interactions create a round inner core: the overall distribution of halo shapes is thus very different and has been used by Peter et al. (2013) to put one of the most stringent constraints on the self-interaction cross-section. However, the presence of baryons can once again reduce this difference at the scale of elliptical galaxies: in Despali et al. (2022), we ran zoom-in hydro simulations in CDM and a SIDM model with $\sigma = 1cm^2g^{-1}$, where galaxy formation is described by the TNG model (see Pillepich et al. (2018)). We calculated intrinsic and projected halo shapes, showing that both self-interactions and baryons contribute to the creation of rounder inner shapes. Overall, the distributions from the two hydro runs are compatible with each other and with the shapes measured in gravitational lensing data - see the right panel of Figure 1).

2.2. *Detection of dark matter subhaloes*

Dark matter subhaloes have been detected via the perturbations that they induce on the surface brightness of lensed arcs Vegetti et al. (2010), Vegetti et al. (2012), Hezaveh et al. (2016). The strength and the shape of the perturbation depend on the density profile of the subhalo: this was originally modelled as a Pseudo-Jaffe profile (with an inner slope equal to -2), while the most recent works use the standard (and shallower) NFW profile. However, the analysis by Minor et al. (2021) has demonstrated that the subhalo concentration must be extremely high to explain the detection in the lens SDSSJ0946+1006 Vegetti et al. (2010), thus questioning the predictions of Cold Dark Matter. Minor et al. (2021) compare the observational results to the structures found in the IllustrisTNG-100 simulation (Pillepich et al. (2018), finding that the simulations do not contain subhaloes that can fully reproduce the data. Nadler, Yang, & Yu (2023) suggest that the tension could be resolved by strong self-interactions that would cause gravothermal core-collapse of some subhaloes and the formation of a very steep density profile.

In a recent work Heinze et al. (2023), we extend the comparison to simulations to the higher resolution TNG50 simulation Pillepich et al. (2019), where the spatial resolution is 288 pc (instead of 740 in TNG100). We find that the subhaloes in TNG50 have profiles that are steeper than the NFW model: the mean inner slope equals -2, and even cuspier systems exist (see Figure 2). The higher spatial resolution allows to trace the density profile more accurately and find candidates that could match the observed values. This is further evidence that baryonic effects need to be taken into account and modelled down to a resolution that is appropriate to each scientific case. In a follow-up paper, we will study the lensing effect of subhaloes in detail and calculate the physical properties that fit the data.

3. Conclusions

In summary, once baryonic physics is properly accounted for, the lensing properties of haloes and subhaloes can be predicted more accurately both in CDM and SIDM scenarios. At the scale of massive elliptical lens galaxies, we conclude that both CDM and a SIDM model with a constant cross-section $\sigma/m_\chi = 1cm^2g^{-1}$ are still compatible with the most recent observations of elliptical galaxies, also consistently with the conclusions of the observational analysis by McDaniel, Jeltema, & Profumo (2021).

Baryonic physics in CDM can also make subhaloes more compact and cuspy, increasing their lensing efficiency. It is possible that the presence of baryons can explain the extreme concentration of the subhalo detected by Vegetti et al. (2010), without invoking alternative SIDM models in which subhaloes undergo gravothermal core-collapse.

It is thus evident how the inclusion of baryons in the simulated results is of fundamental importance to derive realistic predictions, both to confirm or challenge the CDM model. We compared observational results to limited samples of simulated ETGs: larger samples of simulations will help us to derive more precise predictions, avoid selection biases, and base the results on complete samples (see also Adhikari et al. (2022) for a review of future prospects in SIDM models). Moreover, a more consistent analysis of the lensing data would eliminate potential sources of bias and strengthen the comparison with future simulations. Upcoming data sets and new simulations will allow the community to make significant progress in studying the nature of dark matter with astrophysical probes.

References

Adhikari S., Banerjee A., Boddy K. K., Cyr-Racine F.-Y., Desmond H., Dvorkin C., Jain B., et al., 2022, arXiv, arXiv:2207.10638. doi:10.48550/arXiv.2207.10638

Bolton A. S., Burles S., Koopmans L. V. E., Treu T., Gavazzi R., Moustakas L. A., Wayth R., et al., 2008, ApJ, 682, 964. doi:10.1086/589327

Buote D. A., Jeltema T. E., Canizares C. R., Garmire G. P., 2002, ApJ, 577, 183.

Despali G., Walls L. G., Vegetti S., Sparre M., Vogelsberger M., Zavala J., 2022, MNRAS, 516, 4543.

Heinze F., Despali G., Klessen R., 2023, submitted to MNRAS.

Hezaveh Y. D., Dalal N., Marrone D. P., Mao Y.-Y., Morningstar W., Wen D., Blandford R. D., et al., 2016, ApJ, 823, 37.

Mastromarino C., Despali G., Moscardini L., Robertson A., Meneghetti M., Maturi M., 2023, MNRAS, 524, 1515.

McDaniel A., Jeltema T., Profumo S., 2021, JCAP, 2021, 020. doi:10.1088/1475-7516/2021/05/020

Minor Q., Gad-Nasr S., Kaplinghat M., Vegetti S., 2021, MNRAS, 507, 1662.

Nadler E. O., Yang D., Yu H.-B., 2023, arXiv, arXiv:2306.01830. doi:10.48550/arXiv.2306.01830

Nightingale J. W., He Q., Cao X., Amvrosiadis A., Etherington A., Frenk C. S., Hayes R. G., et al., 2022, arXiv, arXiv:2209.10566.

Peter A. H. G., Rocha M., Bullock J. S., Kaplinghat M., 2013, MNRAS, 430, 105.

Pillepich A., Springel V., Nelson D., Genel S., Naiman J., Pakmor R., Hernquist L., et al., 2018, MNRAS, 473, 4077. doi:10.1093/mnras/stx2656

Pillepich A., Nelson D., Springel V., Pakmor R., Torrey P., Weinberger R., Vogelsberger M., et al., 2019, MNRAS, 490, 3196. doi:10.1093/mnras/stz2338

Robertson A., Massey R., Eke V., Schaye J., Theuns T., 2021, MNRAS, 501, 4610.

Sonnenfeld A., Gavazzi R., Suyu S. H., Treu T., Marshall P. J., 2013, ApJ, 777, 97. doi:10.1088/0004-637X/777/2/97

Vegetti S., Koopmans L. V. E., Bolton A., Treu T., Gavazzi R., 2010, MNRAS, 408, 1969.

Vegetti S., Lagattuta D. J., McKean J. P., Auger M. W., Fassnacht C. D., Koopmans L. V. E., 2012, Natur, 481, 341.

Strong Gravitational Lensing in the Era of Big Data
Proceedings IAU Symposium No. 381, 2024
H. Stacey, A. Sonnenfeld & C. Grillo, eds.
doi:10.1017/S1743921323004106

Lensed radio arcs at milli-arcsecond resolution: Methods, science results, and current status

Devon M. Powell

Max Planck Institute for Astrophysics. email: dmpowell@mpa-garching.mpg.de

Abstract. Strong gravitational lensing by galaxies provides us with a powerful laboratory for testing dark matter models. Various particle models for dark matter give rise to different small-scale distributions of mass in the lens galaxy, which can be differentiated with sensitive observations. Th. The sensitivity of a gravitational lens observation to the presence (or absence) of low-mass dark structures in the lens galaxy is determined mainly by the angular resolution of the instrument and the spatial structure of the source. Here, I discuss results from the analysis of a global VLBI observation of a gravitationally lensed radio jet. With an angular resolution better than 5 milli-arcseconds and a highly extended, spatially resolved source, we are able to place competitive constraints on the particle mass in fuzzy dark matter models using this single observation. I also discuss preliminary results from our analysis of warm dark matter models using this lens system.

Keywords. gravitational lensing – dark matter – data analysis – radio continuum – quasars – galaxies

---

1. Introduction

Strong gravitational lensing by galaxies is a powerful observational tool for constraining the particle nature of dark matter. By modeling the distribution of mass inside a lens galaxy (or a sample of lens galaxies) given the observed lensed images, it is possible to rule out dark matter models based on their expected astrophysical phenomenology. For instance, the fiducial cold dark matter (CDM) paradigm predicts that galaxies will contain large populations of gravitationally-bound subhaloes even down to Earth-mass scales (Angulo et al. 2017). By contrast, warm dark matter (WDM) models predict a cutoff in the halo mass function, strongly suppressing the halo population below a certain mass, depending on the mass of the dark matter particle, m_χ (Benson et al. 2013, e.g.). A more exotic model of interest is fuzzy dark matter (FDM), consisting of an ultra-light particle $\mathcal{O}(10^{-21}$ eV), that produces kiloparsec-scale wave-interference structures in the lens galaxy (Schive et al. 2014). Each of these models (we discuss CDM, WDM, FDM here, though there are several other candidates) predicts different small-scale distributions of dark matter in a lens galaxy.

Gravitational lens observations used to test dark matter models fall broadly into two categories. The first are the unresolved lensed quasars, which are lens systems containing quadruply-imaged point sources. The relative positions and fluxes of the images reveal information about the underlying gravitational landscape. Dark matter inferences using lensed quasars require samples of many lenses in order to increase their statistical power (Gilman et al. 2017, 2020; Laroche et al. 2022, e.g.). The second are lens observations containing resolved images and arcs. Such systems contain more information in principle,

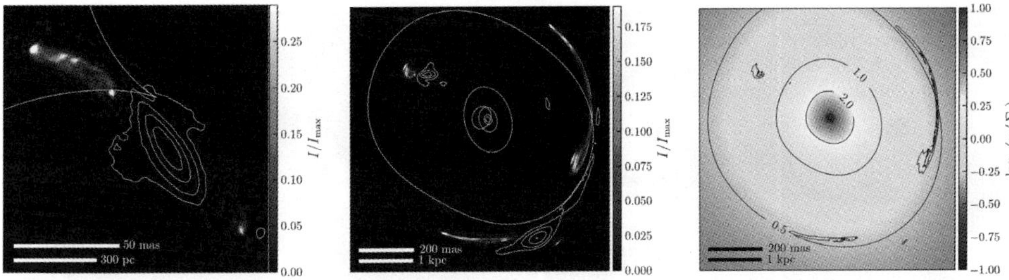

Figure 1. The best composite (baryons plus dark matter) model for MG J0751+2716 from Powell et al. (2022). The continuum radio emission appears in the violet color map, while yellow contours show the infrared emission observed with Keck AO. Critical curves and caustics are shown in white.

but the analysis is more complicated. Individual detections of low-mass dark structures have so far been made in several resolved observations of lens systems observed at angular resolutions $\gtrsim 30$ mas (Vegetti et al. 2010, 2012; Hezaveh et al. 2016b).

In this proceeding, I focus on the case of galaxy-scale lenses containing *resolved* images, specifically those observed at ~ 5 milli-arcsecond resolution using very long baseline interferometry (VLBI). Resolving long, thin arcs at such high resultion can probe mass scales down to $\sim 10^6$ $M_\odot$. In previous work, I presented a computational method for forward-modeling VLBI lens observations directly in the visibility plane (Powell et al. 2021). The 1.6 GHz VLBI data were observed by John McKean and reduced by Spingola et al. (2018). In the rest of this paper, I discuss scientific results to date obtained using these observations and methods.

2. Gravitational lenses are more complex than elliptical power-laws

The first step in modeling a gravitational lens system is to fit a "macro-model", a large-scale parametric mass profile. In past work where lens systems have been observed at relatively low angular resolution ($\gtrsim 70$ mas), macro-models consisting of simple elliptical power-law profiles have been employed successfully.

However, it is becoming increasingly apparent that lens models should include azimuthal degrees of freedom as well. Overly simplistic models can bias inferences on dark matter subhalo populations, as disks and other baryonic structures can be mistaken for dark matter if not properly accounted for (Gilman et al. 2017; Hsueh et al. 2018; He et al. 2022). Contraints on H_0 using time delays between lensed images can also suffer from biases if angular structure is not taken into account (Kochanek 2021; Cao et al. 2022; Van de Vyvere et al. 2022).

In Powell et al. (2022), we found that the VLBI observation of MG J0751+2716 could not be fit by a simple power-law alone, but required a macro-model with angular structure in the form of multipoles (see Figure 1). This result showcased the power of VLBI strong lens observations in probing higher-order complexity in the lens galaxy. This smooth composite model of MG J0751+2716 from Powell et al. (2022) fits the data remarkably well on its own (Figure 1); we use it as the fiducial model for dark matter inferences. However, as I will discuss in following sections, there still exists ambiguity between multipole structure in the lens and the effect of the subhalo population around the lens galaxy.

3. Detecting low-mass dark structures in lens observations

The first application of the MG J0751+2716 VLBI observation and computational method of Powell et al. (2021) to constraining dark matter models was for the case of

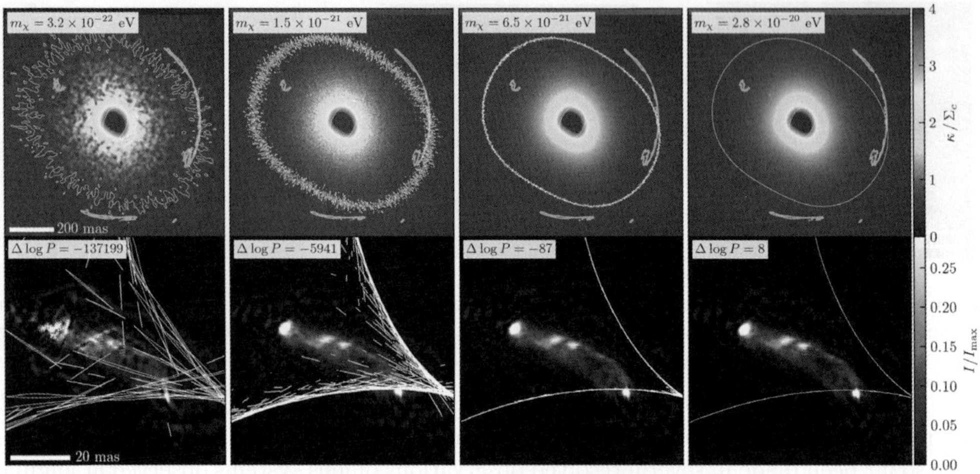

Figure 2. Surface mass density maps with the model lensed images in orange contours (top row) and the corresponding reconstructed source surface brightness maps (bottom row) a sample of random realizations of MG J0751+2716 in an FDM cosmology. Critical curves and caustics are plotted in white. The lensing effect of the FDM granules is apparent: the morphology of the inferred source is disrupted as the model attempts to fit the observation to an incorrect lens model.

fuzzy dark matter (FDM), published in Powell et al. (2023). The main phenomenological feature of galaxy-scale FDM haloes is the presence of wave-interference structures known as "granules" (Schive et al. 2014). The size and density contrast of these granules is determined by the particle mass m_χ.

In Powell et al. (2023), we generated thousands of sample lens realizations in an FDM cosmology and quantified their compatibility with the observed data using their relative log-likelihoods (Figure 2). Extremely light particles produce a very lumpy mass distribution in the lens, and the reconstructed source image is severely disrupted in an attempt to fit the observed data. This is strongly penalized in the log-likelihood. Quantitatively, we ruled out $m_\chi \leq 4.4 \times 10^{-21}$ eV with a 20:1 posterior odds ratio (POR) relative to the fiducial smooth model.

Inferring constraints on the FDM particle is more straightforward than for WDM/CDM for several reasons. First, it is a one-parameter model (m_χ) with relatively simple physics (a classical wave in a potential well). Second, the granule perturbations are non-local and must necessarily appear all along the gravitationally lensed arcs. Third, in the mass range of $m_\chi \sim \times 10^{-21}$ eV, the subhalo population is virtually non-existent. These factors drastically simplify the sampling procedure relative to what is needed to make robust inferences for WDM/CDM halo populations.

In the WDM/CDM case, we are interested in measuring properties of the low-mass halo population, rather than looking for wave-interference granules. WDM haloes have a much more complex formation history (Benson et al. 2013; Ludlow et al. 2016, e.g.), and their gravitational effects are much more localized than FDM granules. There are two main consequences of this localized gravitational effect. The first is that, due to more concentrated objects having a higher lensing efficiency, the detection of individual low-mass haloes with a high concentration is more likely and may bias inferences on population characteristics. The second is that in order for an individual detection to occur, the halo must lie directly on top of a lensed image.

This is problematic for VLBI lenses, as the arcs are typically only a few mas wide, so the sensitive area is very low. In Figure 4, we show a test on a mock observation

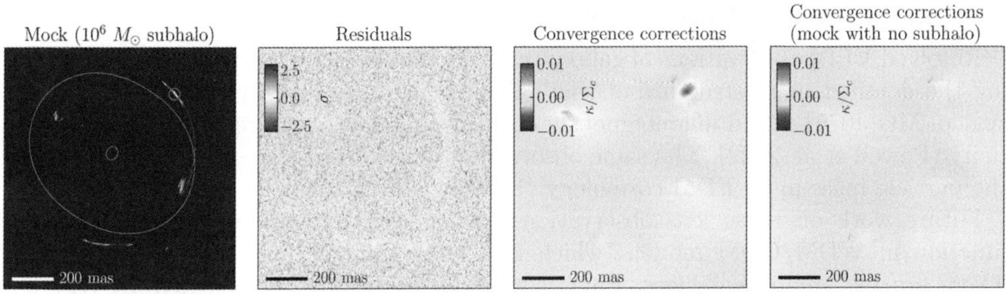

Figure 3. A simulated observation of MG J0751+2716 with a 10^6 $M_\odot$ subhalo on the arc (red circle in the leftmost panel). The subhalo is easily detected via gravitational imaging (second panel from right). While isolated low-mass subhaloes lying on a lensed arc can be individually detected with VLBI observations of this quality, robustly inferring properties of the population way from the lensed arcs remains challenging.

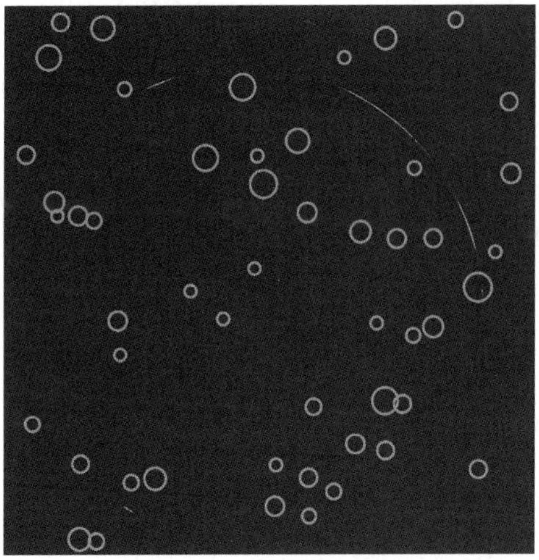

Figure 4. A simulated VLBI observation of the gravitational lens system B1938+666, with a population of dark matter subhaloes in the lens plane (orange circles; the size of each circle corresponds to the mass of the subhalo). The lensed arcs remain smooth in the presence of a collective gravitational perturbation by the subhalo population. (Credit: Simona Vegetti)

of MG J0751+2716 in which a 10^6 $M_\odot$ lies on top of the arc. Using the gravitational imaging (Vegetti and Koopmans 2009), we easily detect its presence. The picture changes, however, when there are no haloes directly on top of the arcs. Figure 3 shows a simulated VLBI observation of another lens system, B1938+666, which also displays long, thin arcs. In this realization, no single subhalo sits directly on an arc. The arcs remain smooth and unperturbed, and the overall effect of the subhalo population is absorbed into the macro-model.

Information on the population of subhaloes and line-of-sight haloes in a VLBI lens observation is contained in some combination of individual detections along the arc, non-detections along the arc (whose statistical significance is tricky to quantify, see e.g. O'Riordan et al. 2023), and smooth perturbations to the arcs caused by the collective off-arc population. While statistical approaches for inferring subhalo population characteristics exist (Vegetti et al. 2014; Hezaveh et al. 2016a), they are more suited to optical/IR observations where the lensed images are spatially extended in two dimensions.

4. Conclusions and future prospects

Resolved VLBI observations of galaxy-scale strong lenses are extremely useful for probing the detailed mass structure of lens galaxies. We have already shown that the lens system MG J0751+2716 unambiguously contains angular structure beyond perfect ellipticity (Powell et al. 2022). The same observation was used to place strong constraints on the particle mass in an FDM cosmology (Powell et al. 2023).

Future work on these lens observations will focus on characterizing the halo mass function in WDM/CDM models, which is a more difficult task than the FDM case. There are two main challenges. First, while long, thin arcs are sensitive to very low-mass dark haloes, their sensitive area is small, so making robust inferences via individual detections or non-detections is difficult. Second, an off-arc population of low-mass haloes has a subtle perturbative effect, shifting the arcs around slightly but leaving them quite smooth. It is thus unclear how well the effects of a sub/LOS-halo population can be distinguished from instrinsic non-ellipticity in the smooth component of the lens galaxy.

We expect sensitivity to low-mass dark structures to scale with angular resolution, so pushing observations to higher frequency (10-100 GHz) will be an important step. The Square Kilometre Array (SKA) will also discover thousands of new radio-bright lenses with extended arcs like this one (McKean et al. 2015). We will continue to develop better modeling approaches to robustly extract as much information as possible from the data. High-resolution VLBI observations are still a relatively new tool in the effort to constrain the particle nature of dark matter with strong lensing. While there is still considerable work to be done on this effort, the future is bright for VLBI and strong lensing.

References

Angulo, R. E., Hahn, O., Ludlow, A. D., & Bonoli, S. 2017, Earth-mass haloes and the emergence of NFW density profiles. *MNRAS*, 471(4), 4687–4701.

Benson, A. J., Farahi, A., Cole, S., Moustakas, L. A., Jenkins, A., Lovell, M., Kennedy, R., Helly, J., & Frenk, C. 2013, Dark matter halo merger histories beyond cold dark matter - I. Methods and application to warm dark matter. *MNRAS*, 428(2), 1774–1789.

Cao, X., Li, R., Nightingale, J. W., Massey, R., Robertson, A., Frenk, C. S., Amvrosiadis, A., Amorisco, N. C., He, Q., Etherington, A., Cole, S., & Zhu, K. 2022, Systematic Errors Induced by the Elliptical Power-law model in Galaxy-Galaxy Strong Lens Modeling. *Research in Astronomy and Astrophysics*, 22(2), 025014.

Gilman, D., Agnello, A., Treu, T., Keeton, C. R., & Nierenberg, A. M. 2017, Strong lensing signatures of luminous structure and substructure in early-type galaxies. *MNRAS*, 467(4), 3970–3992.

Gilman, D., Birrer, S., Nierenberg, A., Treu, T., Du, X., & Benson, A. 2020, Warm dark matter chills out: constraints on the halo mass function and the free-streaming length of dark matter with eight quadruple-image strong gravitational lenses. *MNRAS*, 491(4), 6077–6101.

He, Q., Nightingale, J., Massey, R., Robertson, A., Amvrosiadis, A., Cole, S., Frenk, C. S., Li, R., Amorisco, N. C., Metcalf, R. B., Cao, X., & Etherington, A. 2022, Testing strong lensing subhalo detection with a cosmological simulation. *arXiv e-prints*, arXiv:2202.10191.

Hezaveh, Y., Dalal, N., Holder, G., Kisner, T., Kuhlen, M., & Perreault Levasseur, L. 2016,a Measuring the power spectrum of dark matter substructure using strong gravitational lensing. *J. Cosmology Astropart. Phys.*, 2016a(11), 048.

Hezaveh, Y. D., Dalal, N., Marrone, D. P., Mao, Y.-Y., Morningstar, W., Wen, D., Blandford, R. D., Carlstrom, J. E., Fassnacht, C. D., Holder, G. P., Kemball, A., Marshall, P. J., Murray, N., Perreault Levasseur, L., Vieira, J. D., & Wechsler, R. H. 2016,b Detection of Lensing Substructure Using ALMA Observations of the Dusty Galaxy SDP.81. *ApJ*, 823b, 37.

Hsueh, J.-W., Despali, G., Vegetti, S., Xu, D., Fassnacht, C. D., & Metcalf, R. B. 2018, Flux-ratio anomalies from discs and other baryonic structures in the Illustris simulation. *MNRAS*, 475(2), 2438–2451.

Kochanek, C. S. 2021, Overconstrained models of time delay lenses redux: how the angular tail wags the radial dog. *MNRAS*, 501(4), 5021–5028.

Laroche, A., Gilman, D., Li, X., Bovy, J., & Du, X. 2022, Quantum fluctuations masquerade as haloes: bounds on ultra-light dark matter from quadruply imaged quasars. *MNRAS*, 517(2), 1867–1883.

Ludlow, A. D., Bose, S., Angulo, R. E., Wang, L., Hellwing, W. A., Navarro, J. F., Cole, S., & Frenk, C. S. 2016, The mass-concentration-redshift relation of cold and warm dark matter haloes. *MNRAS*, 460(2), 1214–1232.

McKean, J., Jackson, N., Vegetti, S., Rybak, M., Serjeant, S., Koopmans, L. V. E., Metcalf, R. B., Fassnacht, C., Marshall, P. J., & Pandey-Pommier, M. Strong Gravitational Lensing with the SKA. In *Advancing Astrophysics with the Square Kilometre Array (AASKA14)* 2015, 84.

O'Riordan, C. M., Despali, G., Vegetti, S., Lovell, M. R., & Moliné, Á. 2023, Sensitivity of strong lensing observations to dark matter substructure: a case study with Euclid. *MNRAS*, 521(2), 2342–2356.

Powell, D., Vegetti, S., McKean, J. P., Spingola, C., Rizzo, F., & Stacey, H. R. 2021, A novel approach to visibility-space modelling of interferometric gravitational lens observations at high angular resolution. *MNRAS*, 501(1), 515–530.

Powell, D. M., Vegetti, S., McKean, J. P., Spingola, C., Stacey, H. R., & Fassnacht, C. D. 2022, A lensed radio jet at milliarcsecond resolution I: Bayesian comparison of parametric lens models. *MNRAS*, 516(2), 1808–1828.

Powell, D. M., Vegetti, S., McKean, J. P., White, S. D. M., Ferreira, E. G. M., May, S., & Spingola, C. 2023, A lensed radio jet at milli-arcsecond resolution - II. Constraints on fuzzy dark matter from an extended gravitational arc. *MNRAS*, 524(1), L84–L88.

Schive, H.-Y., Liao, M.-H., Woo, T.-P., Wong, S.-K., Chiueh, T., Broadhurst, T., & Hwang, W. Y. P. 2014, Understanding the Core-Halo Relation of Quantum Wave Dark Matter from 3D Simulations. *Phys. Rev. Lett.*, 113(26), 261302.

Spingola, C., McKean, J. P., Auger, M. W., Fassnacht, C. D., Koopmans, L. V. E., Lagattuta, D. J., & Vegetti, S. 2018, SHARP - V. Modelling gravitationally lensed radio arcs imaged with global VLBI observations. *MNRAS*, 478(4), 4816–4829.

Van de Vyvere, L., Gomer, M. R., Sluse, D., Xu, D., Birrer, S., Galan, A., & Vernardos, G. 2022, TDCOSMO. VII. Boxyness/discyness in lensing galaxies: Detectability and impact on H_0. *A&A*, 659, A127.

Vegetti, S. & Koopmans, L. V. E. 2009, Bayesian strong gravitational-lens modelling on adaptive grids: objective detection of mass substructure in Galaxies. *MNRAS*, 392(3), 945–963.

Vegetti, S., Koopmans, L. V. E., Auger, M. W., Treu, T., & Bolton, A. S. 2014, Inference of the cold dark matter substructure mass function at z = 0.2 using strong gravitational lenses. *MNRAS*, 442(3), 2017–2035.

Vegetti, S., Koopmans, L. V. E., Bolton, A., Treu, T., & Gavazzi, R. 2010, Detection of a dark substructure through gravitational imaging. *MNRAS*, 408(4), 1969–1981.

Vegetti, S., Lagattuta, D. J., McKean, J. P., Auger, M. W., Fassnacht, C. D., & Koopmans, L. V. E. 2012, Gravitational detection of a low-mass dark satellite galaxy at cosmological distance. *Nature*, 481(7381), 341–343.

Strong Gravitational Lensing in the Era of Big Data
Proceedings IAU Symposium No. 381, 2024
H. Stacey, A. Sonnenfeld & C. Grillo, eds.
doi:10.1017/S1743921323004143

Clumpiness of lens galaxies as a window on dark matter

Dorota Bayer

Centre for Astrophysics & Supercomputing, Swinburne University of Technology, Hawthorn, Australia
ARC Centre of Excellence for All Sky Astrophysics in 3 Dimensions (ASTRO 3D), Australia.
email: dorota.bayer@gmx.de

Abstract. While the direct detection of the dark-matter particle remains very challenging, the physical properties of dark matter could potentially be constrained indirectly – by comparing the population characteristics of substructures in real galaxies with predictions from the phenomenological dark-matter models, such as cold, warm or self-interacting dark matter. Whereas these models are practically indistinguishable with respect to the expected abundance and statistical properties of massive galactic substructures, the critical difference lies in the low-mass regime $\lesssim 10^8 M_\odot$. Galaxy-galaxy strong gravitational lensing provides a unique opportunity to search for gravitational signatures of such low-mass substructures in galaxies acting as a strong gravitational lens on a more distant background galaxy, serendipitously located along the same line of sight. In Bayer et al. (2023a,b,c), we have recently introduced a novel approach to investigate the collective perturbative effect of low-mass substructures in the inner regions of massive elliptical lens galaxies and observationally constrain their power spectrum (on 1-10 kpc scales) based on the Fourier analysis of the associated surface-brightness anomalies in the lensed images (i.e., Einstein rings and gravitational arcs) observed with the Hubble Space Telescope. Here, we present a concise summary of the methodology, the encountered modelling challenges and the inferred observational constraints on the sub-galactic matter power spectrum. The future comparison of these results with predictions from hydrodynamical simulations might either verify the cold-dark-matter paradigm or require its substantial revision. Moreover, we demonstrate that the grid-based smooth lens modelling might model away surface-brightness anomalies caused by the presence of substructures in the lens galaxy and incorrectly interpret them as surface-brightness structure in the lensed background galaxy itself. If not properly understood and accounted for, such signal suppression might lead to severely biased constraints on the properties of substructures in galactic haloes.

Keywords. Cosmology, Dark matter, Galactic substructure, Gravitational lensing

1. Introduction

One of the major unsolved puzzles in cosmology and the fundamental physics is the dark-matter problem – despite the essential role that dark matter plays in the cosmological structure-formation process, its nature and properties remain unknown. The standard cold dark matter (CDM) paradigm is still challenged by various alternative models, such as warm dark matter (WDM) or self-interacting dark matter (SIDM). While the direct detection of the dark-matter particle continues to pose a challenge, one of the most promising avenues to unveil the nature of dark matter is the study of the abundance and statistical properties of dark-matter substructures in galactic haloes. The competing dark-matter models, such as CDM, WDM or SIDM, lead to significantly different levels of mass clumpiness/smoothness on the sub-galactic scales. Whereas the CDM model predicts galactic haloes to be inhabited by an abundant population of substructures on mass scales down to at least $10^{-6} M_{\odot}$, models with less massive or self-interacting dark-matter particle significantly suppress the formation of substructures, especially in the low-mass regime below $\approx 10^{8} M_{\odot}$.

Galaxy-galaxy strong gravitational lensing offers a unique opportunity to search for gravitational signatures of such low-mass dark-matter substructures in lens galaxies beyond the Local Group. Substructures in the lens galaxy (or along its line-of-sight) induce perturbations to the otherwise smoothly-distributed lensing potential. The resulting observable *surface-brightness anomalies* in the extended lensed images of the background galaxy (Einstein rings or gravitational arcs), i.e., deviations of the observed surface-brightness distribution from the prediction of a smooth-lens model, can be modelled and traced back to the underlying substructure in the lens galaxy. One of the most successful approaches in this respect is the *gravitational-imaging technique* (Koopmans 2005; Vegetti and Koopmans 2009), which allows one to constrain the projected position and mass of individual massive subhaloes in (massive elliptical) lens galaxies via their imprints on the lensed images. However, the current mass-detection threshold of this technique ($\sim 10^{8} M_{\odot}$ for HST-observations of SLACS lenses) still lies above the mass regime in which the alternative dark-matter models could be distinguished, given the current sample sizes and data sensitivity.

In response to these observational limitations, in Bayer et al. (2023a,b,c), we have recently introduced an alternative statistical approach. This aims at investigating the overall clumpiness in the total (dark and baryonic) mass distribution of massive elliptical lens galaxies, i.e., the level of small-scale deviations from the best-fitting smooth elliptical power-law mass density model with external shear, and inferring observational constraints on the *sub-galactic matter power spectrum* from *the power spectrum of the collectively-induced surface-brightness anomalies* measured in highly magnified galaxy-scale Einstein rings and gravitational arcs. Here, we present a concise summary of the methodology, the encountered modelling challenges and the most important results.

2. Methodology

In Bayer et al. (2023a), we have presented a methodology allowing us to reliably measure the power spectrum of surface-brightness anomalies in extended lensed images of galaxy-galaxy strong gravitational lens systems (i.e., Einstein rings or gravitational arcs), observed with the Hubble Space Telescope (HST/WFC3/F390W), through the following data-analysis steps:

(1) modelling and subtruction of the lens-galaxy light;
(2) simultaneous reconstruction of the best-fitting smoothly-varying lensing potential and the intrinsic unlensed surface-brightness distribution of the background galaxy;
(3) statistical quantification of the residual surface-brightness fluctuations in the lensed images in terms of the azimuthally-averaged power spectrum $P_{\delta I}(k)$;
(4) estimation and correction for the noise power spectrum in the HST-observations.

The noise correction of the residual surface-brightness fluctuations requires a particularly careful consideration due to the noise correlations caused by the combined effect of the drizzling procedure and the charge transfer inefficiency of the HST/WFC3/UVIS-CCDs. To this end, we have developed a methodology to estimate the noise power spectrum in the analysed science images based on a set of blank-sky cutouts from the full field of view that we overlay with the Poisson noise of the specific observation.

In Bayer et al. (2023b), we have extended this methodology to trace the revealed surface-brightness anomalies back to the underlying population of small-scale density variations in the lens galaxy, which can arise not only from the low-mass dark-matter substructures and line-of-sight haloes but also from density inhomogeneities in the baryonic mass distribution, such as e.g. globular clusters or tidal streams. As proposed by Chatterjee and Koopmans (2018), we model the overall density variations as Gaussian Random Field (GRF) potential perturbations $\delta\psi_{\mathrm{GRF}}(\boldsymbol{x})$ superposed on a smoothly-varying lensing potential and assume them to be uniquely characterised by a power-law power spectrum $P_{\delta\psi}(\sigma^2_{\delta\psi}, \beta; k)$ with two free parameters: the total variance $\sigma^2_{\delta\psi}$ and the power-spectrum slope β. The latter describes the distribution of the integrated variance over the different length scales (i.e., k-modes) and, thus, quantifies the scale dependence of the hypothetical small-scale density variations.

To interpret the real measurement, we have performed a systematic study of simulated surface-brightness anomalies induced by low-mass substructures with varying statistical properties. The investigation is performed in the parameter space spanned by $\sigma^2_{\delta\psi}$ and β. For each point on this grid, referred to as a *matter-power spectrum model*, we have generated realizations of Gaussian potential perturbations, superposed them on the best-fitting smooth lensing potential inferred for the real data, computed the resulting perturbed lensed images and, finally, determined the power spectrum of the induced surface-brightness anomalies. This catalogue of mock power spectra can be directly compared to the actual measurement, which allows us to determine exclusion probabilities for all considered matter-power spectrum models, as demonstrated in Fig. 1. A valuable insight for future studies is that the fundamental quantity determining the exclusion probability of the matter-power-spectrum models is the integrated variance of the perturbations in the deflection angle $\sigma^2_{\delta\alpha}$ resulting from the induced potential perturbations $\delta\psi_{\mathrm{GRF}}(\boldsymbol{x})$. The exclusion probability is almost insensitive to the slope of the power spectrum of the deflection angles $P_{\delta\alpha}(k)$.

3. Observational constraints on the sub-galactic matter power spectrum

The pilot application of the complete methodology to HST/WFC3/F390W-observations of the lens system SDSS J0252+0039 from the SLACS Survey has allowed us to infer the first observational constraints on the power spectrum of low-mass sub-galactic structures on 1-10 kpc scales, as is presented in Fig. 1. Based on this pilot study, we have ruled out (with the exclusion probability of at least 99 per cent) potential perturbations $\delta\psi_{\mathrm{GRF}}(\boldsymbol{x})$ with the variance $\sigma^2_{\delta\psi}$ exceeding $\sim 10^{-2.5}$ in the analysed field of view of ≈ 20 kpc on a side for a wide range of power-spectrum slopes β. In order to make these results less dependent on the chosen field of view, we have also derived the corresponding constraints on the dimensionless convergence power spectrum $\Delta^2_{\delta\kappa}(\lambda)$ for three different sub-galactic scales. Accordingly, we rule out matter-power-spectrum models with $\Delta^2_{\delta\kappa}(0.5\ \mathrm{kpc}) > 1$ on 0.5-kpc scale, $\Delta^2_{\delta\kappa}(1\ \mathrm{kpc}) > 0.1$ on 1-kpc scale and $\Delta^2_{\delta\kappa}(3\ \mathrm{kpc}) > 0.01$ on 3-kpc scale at the 99 per cent confidence level. Alternatively, these constraints can be also translated into upper limits on the standard deviation of the aperture mass (i.e., the total mass within a cylinder with the diameter λ projected onto the lens plane) in the proximity to the Einstein radius of the lens galaxy: $\sigma_{AM}(0.5\ \mathrm{kpc}) < 0.8 \times 10^8 M_\odot$ inside the circular aperture of diameter $\lambda = 0.5$ kpc, $\sigma_{AM}(1\ \mathrm{kpc}) < 1 \times 10^8 M_\odot$ for $\lambda = 1$ kpc and $\sigma_{AM}(3\ \mathrm{kpc}) < 3 \times 10^8 M_\odot$ for $\lambda = 3$ kpc. Due to the encountered degeneracy between the real surface-brightness anomalies induced by mass structure in the lens galaxy and the complex intrinsic surface-brightness distribution of the star-forming background galaxy itself, the inferred constraints can be expressed only in terms of an upper limit.

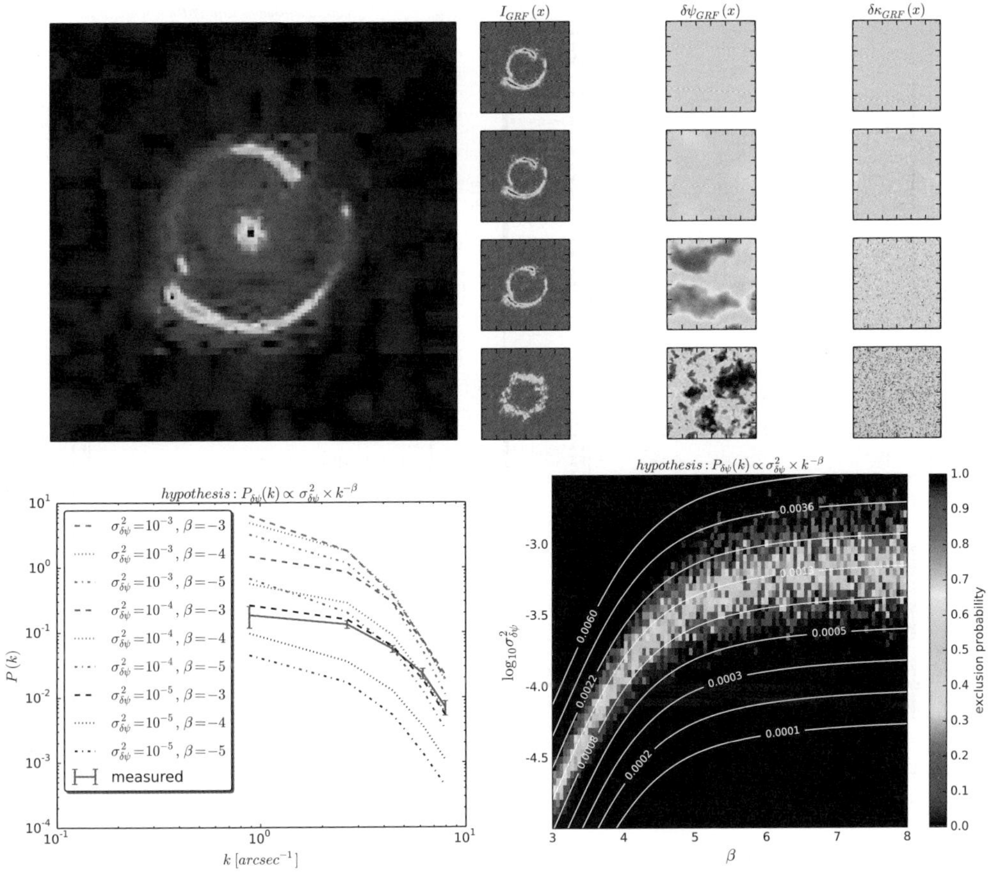

Figure 1. Upper-limit constraints on the sub-galactic matter power spectrum in the massive elliptical lens galaxy SDSS J0252+0039. *Top left panel*: HST/WFC3/F390W imaging of the strong gravitational lens system SDSS J0252+0039 with the side length of 4.48 arcsec. *Top right panel*: Mock surface-brightness anomalies induced in the lensed images by Gaussian potential perturbations $\delta\psi_{\mathrm{GRF}}(\boldsymbol{x})$ in the lens galaxy. These are assumed to be fully characterized by the power spectrum $P_{\delta\psi}(k) \propto \sigma^2_{\delta\psi} \times k^{-\beta}$ with the integrated variance $\sigma^2_{\delta\psi}$ and the power-law slope β. *Bottom left panel*: The power spectrum of surface-brightness anomalies measured in real observations *(red line)* in comparison to the power spectra of mock surface-brightness anomalies induced by Gaussian potential perturbations with specific values of $\sigma^2_{\delta\psi}$ and β *(dashed lines)*. *Bottom right panel*: Exclusion probabilities for the integrated variance $\sigma^2_{\delta\psi}$ and the power-law slope β. The *white contour lines* connect points corresponding to the same variance of the resulting deflection-angle perturbations $\sigma^2_{\delta\alpha}$ in the analysed field of view; based on Bayer et al. (2023a,b).

4. Signal suppression in grid-based smooth lens modelling

A critical issue in our analysis has been the degeneracy between the real surface-brightness anomalies due to mass structure in the lens galaxy and the complex intrinsic surface-brightness distribution of the star-forming background galaxy itself. Grid-based source modelling can model away some of the anomalies caused by mass structure in the lens galaxy and partially even the observational noise. Due to numerical effects, this signal suppression is particularly severe when the lens modelling is performed in a narrow field of view with the highest source-grid resolution. A practical solution to this

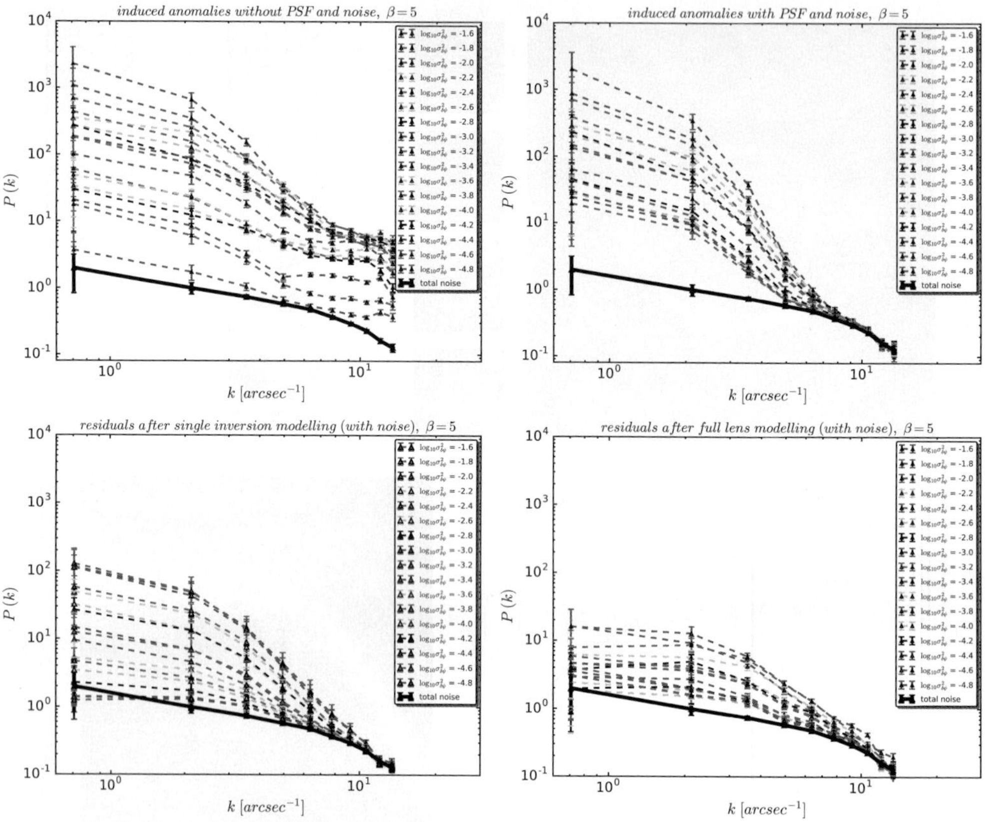

Figure 2. Signal suppression in grid-based smooth lens modelling. _Top_: Power spectra of mock surface-brightness anomalies induced in the lensed images of SDSS J0946+1006 by GRF-potential perturbations characterised by the power spectrum $P_{\delta\psi}(k) \propto \sigma^2_{\delta\psi} \times k^{-\beta}$ with different levels of the total variance $\sigma^2_{\delta\psi}$ and a fixed power-spectrum slope $\beta = 5$ before *(left panel)* vs after *(right panel)* accounting for the observational effects (i.e., convolution with the PSF and adding a realistic noise realization) for a set of 17 mocks. _Bottom_: Suppressed power spectra of mock surface-brightness anomalies after a single-inversion reconstruction of the source galaxy with no optimisation of the lensing potential *(left panel)* vs the *full* grid-based smooth lens modelling of the mock lensed images, i.e., simultaneous optimization of the lensing potential and reconstruction of the background source on a grid *(right panel)*; based on Bayer et al. (2023c).

problem is to lower the resolution of the source reconstruction, while keeping fixed the best-fitting parameter values of the lensing potential obtained with the highest resolution. Alternatively, this problem can be also partly alleviated by the choice of a larger mask covering not only the lensed images but also additional pixels located close to the edge of the analysed cutout. These carry additional information about the noise level in the analysed image and provide more constraints for the Bayesian lens-modelling procedure, which improves the quality of the resulting lens models.

Our most recent work (Bayer et al. 2023c) is fully devoted to alleviating this degeneracy and paving the way for a future unbiased *detection*. With this goal in mind, we propose and test a more accurate version of the methodology to model the simulated perturbed lensed images in exactly the same manner as the real data. The modification is introduced gradually in two steps. First, we perform the less computationally expensive single-inversion modelling (i.e., reconstruction of the pixelized surface-brightness distribution of the background galaxy, while keeping fixed the parameter values of the lensing potential obtained with the highest resolution), which is suitable for studies of large samples. Second, we carry out the much more accurate full grid-based smooth lens modelling (i.e., simultaneous modelling of the lens and source galaxy). The latter approach allows us to compare the observed and the simulated lensed images in a fully consistent manner, however, it is significantly more time-consuming and introduces another layer of complexity and modelling choices to be made.

To quantify the effect of signal suppression and test the feasibility of the improved methodology, we apply it to deep high-signal-to-noise HST/ACS/F814W-observations of the lens system SDSS J0946+1006, commonly known as the Double Einstein Ring, and a set of 17 mock perturbed lensed images mimicking SDSS J0946+1006. The results presented in Fig. 2 show a striking suppression of the true surface-brightness anomalies in the course of the full grid-based smooth lens modelling (up to ~2 orders of magnitude for the most perturbed lens systems), especially for the lowest considered k-values (i.e., largest scales). This finding implies that the surface-brightness anomalies measured in the real data might be significantly suppressed as well and the analysed lens system could in reality be considerably more perturbed. Therefore we conclude that signal suppression in the grid-based smooth lens modelling is an important issue that requires more attention from the lensing community. If not properly understood and accounted for, it might lead to severely biased results with respect to both the level of substructure in lens galaxies and the properties of lensed high-redshift sources.

References

D. Bayer, S. Chatterjee, L. V. E. Koopmans, S. Vegetti, J. P. McKean, T. Treu, C. D. Fassnacht, and K. Glazebrook. Probing sub-galactic mass structure with the power spectrum of surface-brightness anomalies in high-resolution observations of galaxy-galaxy strong gravitational lenses. II. Observational constraints on the subgalactic matter power spectrum. *MNRAS*, 523(1):1310–1325, July 2023a. doi: 10.1093/mnras/stad1402.

D. Bayer, L. V. E. Koopmans, J. P. McKean, S. Vegetti, T. Treu, C. D. Fassnacht, and K. Glazebrook. Probing sub-galactic mass structure with the power spectrum of surface-brightness anomalies in high-resolution observations of galaxy-galaxy strong gravitational lenses - I. Power-spectrum measurement and feasibility study. *MNRAS*, 523(1):1326–1345, July 2023b. doi: 10.1093/mnras/stad1403.

D. Bayer, G. Vernardos, and L. V. E. Koopmans. Probing sub-galactic mass structure with the power spectrum of surface-brightness anomalies in high-resolution observations of galaxy-galaxy strong gravitational lenses - III. Suppression of surface-brightness anomalies in grid-based lens modelling. *MNRAS*, 2023c. doi: submitted.

S. Chatterjee and L. V. E. Koopmans. The inner mass power spectrum of galaxies using strong gravitational lensing: beyond linear approximation. *MNRAS*, 474:1762–1772, February 2018. doi: 10.1093/mnras/stx2674.

L. V. E. Koopmans. Gravitational imaging of cold dark matter substructures. *MNRAS*, 363: 1136–1144, November 2005. doi: 10.1111/j.1365-2966.2005.09523.x.

S. Vegetti and L. V. E. Koopmans. Bayesian strong gravitational-lens modelling on adaptive grids: objective detection of mass substructure in Galaxies. *MNRAS*, 392:945–963, January 2009. doi: 10.1111/j.1365-2966.2008.14005.x.

Strong Gravitational Lensing in the Era of Big Data
Proceedings IAU Symposium No. 381, 2024
H. Stacey, A. Sonnenfeld & C. Grillo, eds.
doi:10.1017/S1743921323004131

Angular structure and gravitational imaging

Conor M. O'Riordan

Max Planck Institut für Astrophysik, Karl-Schwarzschild-Straße 1, 85748 Garching bei München, Germany. email: conor@mpa-garching.mpg.de

Abstract. In gravitational imaging, the mass model for the main lensing galaxy is one of the main sources of systematic uncertainty. We use subhalo detection models with increasing levels of angular complexity in the lens mass model to analyse 100 HST mock observations. We find that perturbations of just 1% are enough to cause a 20% false positive subhalo detection rate, with order 3 multipoles having the strongest effect. The area in an observation where a substructure can be detected drops by a factor of 10 if multipoles up to 3 per cent amplitude are included in the lens model. The mass of the smallest detectable substructure however is not affected. We find a detection limit of $M > 10^{8.2} M_\odot$ at 5σ in all models. In order for strong lensing searches for dark matter objects to remain reliable in the future, angular structure beyond the elliptical power-law must be included.

Keywords. gravitational lensing, dark matter, machine learning

---

1. Introduction

Strong gravitational lensing has become one of the most important tools for investigating the nature of dark matter (Vegetti et al. 2023). This is because it can be used to detect dark matter subhaloes in the environments of galaxies. The existence of a large number of these subhaloes is a key prediction of the most popular dark matter model, cold dark matter (CDM) (Springel et al. 2008). With a technique called gravitational imaging, the number and mass of these subhaloes can be measured in strong lenses, constraining the underlying DM model (Vegetti and Koopmans 2009).

Gravitational imaging however is an expensive method. This is mostly due to the final stage of the analysis: so-called sensitivity mapping. Here, the observation is analysed to find the smallest detectable subhalo in each pixel (Despali et al. 2022). This information can be used to turn a set of subhalo detections and non-detections into an inference on the dark matter model. In O'Riordan et al. (2023) we introduced a machine learning technique that uses a set of large convolutional neural networks (CNNs) to replace the expensive sensitivity mapping stage. We exploited this new technique to test the sensitivity of *Euclid* strong lenses to dark matter subhaloes. Analysing 16,000 simulated *Euclid* strong lens observations we found that subhaloes with mass larger than $M = 10^{8.8\pm0.2} M_\odot$ could be detected at 3σ in that data, and that the entire survey should yield ~ 2500 new detections.

In the current work, we take our method much further to understand a crucial systematic uncertainty in subhalo detection: the angular complexity in the lens mass model. Strong lenses are typically modelled as elliptical power-laws, or some variation thereof. In this case the lens mass model has a strict elliptical symmetry, although it is often perturbed by an external shear component. We consider multipole perturbations to this simple model, where Fourier modes of order m are added to the projected density of the

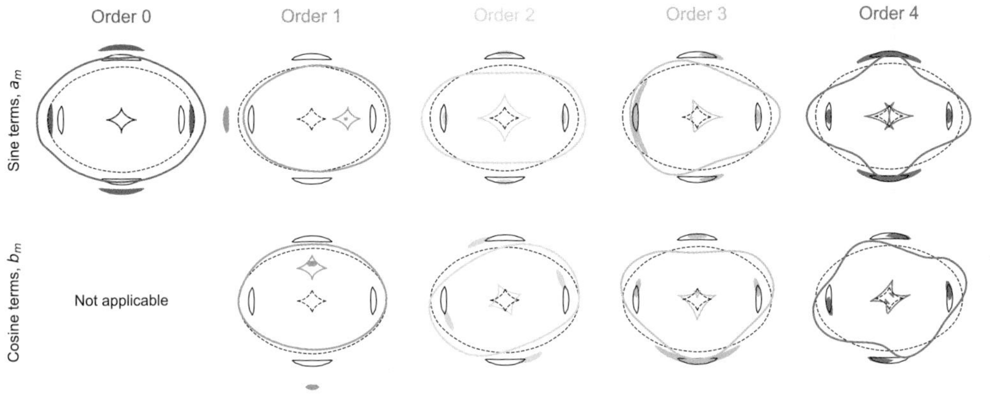

Figure 1. The effect on the critical curves, lensed images, and caustics from multipole perturbations. The macro model is an isothermal ellipsoid with axis ratio $q = 0.7$ and a source at the origin. The unperturbed critical curves and caustics are plotted as dashed black lines. The perturbed versions are plotted as solid coloured lines. In each case the perturbation is 0.1 in the respective coefficient.

galaxy. We use the definition given by Powell et al. (2022) for the convergence κ_m due to a multipole of order m,

$$\kappa_m(\theta, \phi) = \theta^{1-\gamma} \left[a_m \sin(m\phi) + b_m \cos(m\phi)\right], \tag{1}$$

where θ is the angular radial distance from the centre of the lens, ϕ is the polar angle, γ is the slope of the lensing galaxy power-law mass profile, and a_m and b_m are the sine and cosine amplitudes. For convenience we also define the multipole strength, η_m, and multipole angle, ϕ_m, as

$$\eta_m^2 = a_m^2 + b_m^2, \tag{2}$$

$$\phi_m = \arctan(b_m/a_m), \tag{3}$$

respectively. Figure 1 shows the effect of these multipole perturbations on the lens. Perturbations of this nature are observed in the isophotes of elliptical galaxies (Bender and Moellenhoff 1987; Hao et al. 2006; Chaware et al. 2014; Mitsuda et al. 2017). Of particular importance is the $m = 4$ perturbation which gives rise to boxiness/disciness. This specific type of structure is motivated by the results of simulations, where it is expected to result from mergers (Naab et al. 1999).

The effect of these perturbations on measuring the Hubble constant in strong lenses has been studied by Van de Vyvere et al. (2022). The effect on substructure detection has not yet been systematically analysed, although multipole perturbations do feature in the lens models of some studies (e.g. Nightingale et al. 2022; Powell et al. 2022). Both of these examples find multipole amplitudes of order 0.01 in fits to real strong lenses.

We show that small changes in the angular structure of the lens galaxy can be easily mistaken for substructures when an insufficiently complex model is used. When a more complicated model is used, the area of the observation in which a substructure can be detected drastically changes, although the smallest detectable mass of substructure does not. We argue therefore that more angular complexity is required in strong lensing models for dark matter constraints to be reliable.

2. Method

We develop three neural networks to detect subhaloes in simulated strong lens images. The images have HST pixel scale and PSF. The source galaxies are taken directly from images of Hubble deep field (HDF) galaxies. Source galaxies are split into a training

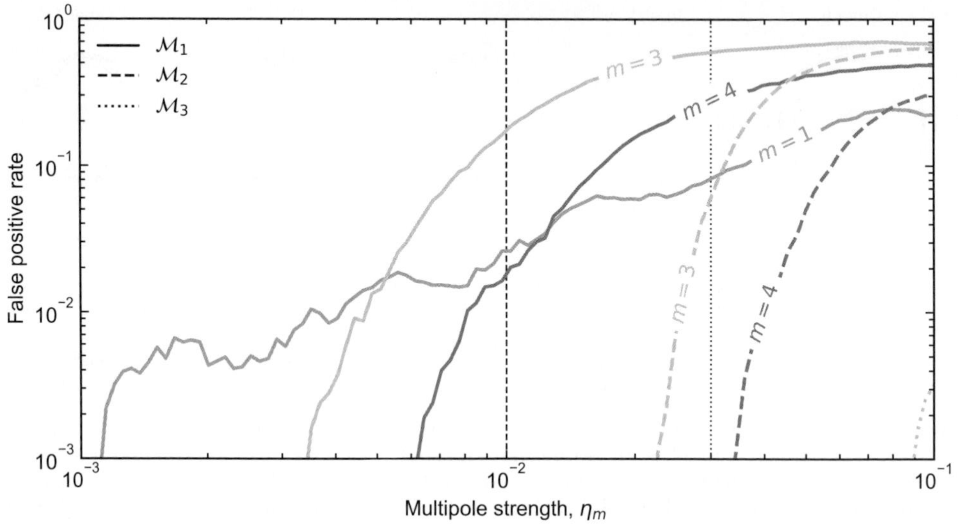

Figure 2. The rate of false positive substructure detection in observations with multipole perturbations but without subhaloes, as a function of multipole strength η_m and order m. The multipole order used is labelled on each curve. The solid curves show the FPR for the model trained with no multipole perturbations.

and testing set. The testing set sources are also used for our results. Lens galaxies are elliptical power-laws. We place the lenses randomly in the volume of the HDF and look for sources in the light cone behind the lens. This produces realistic distributions of lens and source redshift, as well as Einstein radius. In producing training data we resample most lens and source parameters to produce as uniform a distribution of observations as possible. In our results, we use a set of 100 systems drawn from the realistic catalogue which resemble SLACS lenses Bolton et al. (2006). These have large, full Einstein rings with high S/N and are the lenses most useful for gravitational imaging.

To train the models we start with a parameter sweep to find the best initial learning rate. We optimise for the cross-entropy loss using the Adam optimiser. During training, when the loss on the testing data does not improve for 10 epochs, the learning rate is decayed by a factor $10^{-0.5}$. When three decays in learning have not improved the testing loss, the model is assumed to have converged. By the end of training, each model has seen $\sim 10^7$ unique strong lens images.

We label the three neural network models M_1, M_2 and M_3. In M_1, the training data only includes elliptical power-law lenses plus external shear up to a strength of $\gamma_{\mathrm{ext}} < 0.1$. In M_2 and M_3, random multipole perturbations of orders 1, 3, and 4 are present in the training data. The multipole strength is $\eta_m < 0.01$ in M_2 and $\eta_m < 0.03$ in M_3.

3. Results

Figure 2 shows the rate of false positive detections for the three trained models. At each multipole strength value we create 100 realisations of each of the 100 HST mock observations, with random multipole angles. These are passed through the detection models and the rate of 5σ detections is recorded. An amplitude of $\eta_m = 0.01$ causes a false positive subhalo detection 2.6 per cent, 18 per cent, and 2.0 per cent of the time for orders 1, 3, and 4 respectively. At $\eta_m = 0.03$ the same rates are 9.4 per cent, 61 per cent, and 30 per cent. At all multipole strengths, the order 3 perturbation is the most effective at causing false positive detections, with order 4 having a similar but weaker effect.

We use the three models to compute sensitivity maps for our 100 mock HST observations. The sensitivity maps give the lowest mass detectable subhalo in each pixel. Figure 3

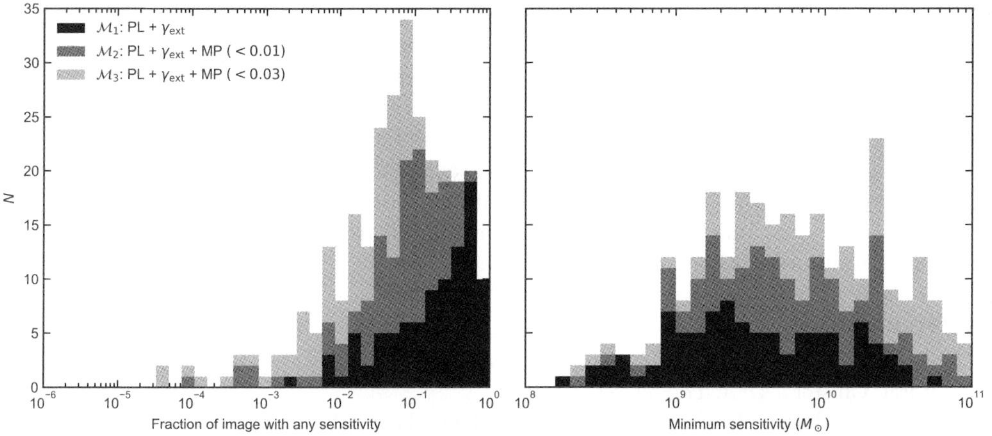

Figure 3. Summary statistics for sensitivity maps produced from our 100 mock HST observations, using all three models. Left: the fraction of the observation which is sensitive to any mass of substructure $M < 10^{11} M_\odot$ at 5σ. Right: the detectable mass in the most sensitive pixel in each observation.

summarises the sensitivity maps in two statistics. On the left of the figure, we plot the total area in each observation which is sensitive to any mass using each model. When no multipoles are included in the model, 29 per cent of the area of all images is sensitive. This drops to 10 per cent when multipoles up to $\eta_m < 0.01$ are included and 3.6 per cent for $\eta_m < 0.03$. In contrast, the lowest detectable mass, plotted on the right of Fig. 3 does not change significantly between the three models. The most sensitive pixel in over all observations loses only 0.11 dex of sensitivity when including multipoles, with no further change when increasing the allowed size of the multipoles.

Sensitivity maps also allow one to calculate the expected number of detectable objects in an observation. A given dark matter model has a specific subhalo mass function, which is the number distribution of subhaloes on the sky as a function of mass. With the lowest detectable mass from the sensitivity map, we can integrate over this function to get the expected number of detectable objects in each pixel and sum over all pixels to get that number for each observation.

For the model with no included multipoles, and assuming CDM, we find 0.048 detectable objects per lens, or one detection in every ~ 20 observations. This is consistent with the frequency of detections in real HST data (Vegetti et al. 2014; Nightingale et al. 2022). When allowing for multipoles up to $\eta_m < 0.01$, this frequency drops to 0.0095 and 0.0047 for multipoles up to $\eta_m < 0.03$, equivalent to a detection once in every ~ 100 and ~ 200 lenses respectively.

4. Conclusions

When modelling gravitational lenses to reveal the presence of dark matter subhaloes, the lens mass model is a significant source of uncertainty. The angular complexity of the mass model is especially important, given that galaxies in nature can deviate significantly from the assumed elliptical symmetry of most analyses. In this work, we systematically analysed the degeneracy between non-elliptical angular structure in the lens galaxy, parameterised with multipole perturbations, and dark matter subhaloes. We found that small deviations from ellipticity, consistent with those found in nature, can cause a significant rate of false positive substructure detection when those deviations are not included in the lens model.

Allowing for deviations from ellipticity in the lens model has a significant effect on the sensitivity of strong lens observations to dark matter substructure, reducing the area in the observation where a subhalo can be detected by up to a factor of 10. The minimum detectable subhalo mass however does not drastically change when including these perturbations. We therefore expect detections of subhaloes at the high-mass end, far from the lensed images to be most effected. These are also the least informative detections in terms of constraining dark matter models. We recommend therefore that future gravitational imaging efforts allow for complicated angular structure in the lens mass model, in order to remain reliable.

References

Bender, R. & Moellenhoff, C. 1987, Morphological analysis of massive early-type galaxies in the Virgo Cluster. *A&A*, 177, 71–83.

Bolton, A. S., Burles, S., Koopmans, L. V. E., Treu, T., & Moustakas, L. A. 2006, The Sloan Lens ACS Survey. I. A Large Spectroscopically Selected Sample of Massive Early-Type Lens Galaxies. *ApJ*, 638(2), 703–724.

Chaware, L., Cannon, R., Kembhavi, A. K., Mahabal, A., & Pandey, S. K. 2014, Isophotal Shapes of Early-type Galaxies to Very Faint Levels. *ApJ*, 787(2), 102.

Despali, G., Vegetti, S., White, S. D. M., Powell, D. M., Stacey, H. R., Fassnacht, C. D., Rizzo, F., & Enzi, W. 2022, Detecting low-mass haloes with strong gravitational lensing I: the effect of data quality and lensing configuration. *MNRAS*, 510(2), 2480–2494.

Hao, C. N., Mao, S., Deng, Z. G., Xia, X. Y., & Wu, H. 2006, Isophotal shapes of elliptical/lenticular galaxies from the Sloan Digital Sky Survey. *MNRAS*, 370(3), 1339–1350.

Mitsuda, K., Doi, M., Morokuma, T., Suzuki, N., Yasuda, N., Perlmutter, S., Aldering, G., & Meyers, J. 2017, Isophote Shapes of Early-type Galaxies in Massive Clusters at z $\sim$ 1 and 0. *ApJ*, 834(2), 109.

Naab, T., Burkert, A., & Hernquist, L. 1999, On the Formation of Boxy and Disky Elliptical Galaxies. *ApJ*, 523(2), L133–L136.

Nightingale, J. W., He, Q., Cao, X., Amvrosiadis, A., Etherington, A., Frenk, C. S., Hayes, R. G., Robertson, A., Cole, S., Lange, S., Li, R., & Massey, R. 2022, Scanning For Dark Matter Subhalos in Hubble Space Telescope Imaging of 54 Strong Lenses. *arXiv e-prints*, arXiv:2209.10566.

O'Riordan, C. M., Despali, G., Vegetti, S., Lovell, M. R., & Moliné, Á. 2023, Sensitivity of strong lensing observations to dark matter substructure: a case study with Euclid. *MNRAS*, 521(2), 2342–2356.

Powell, D. M., Vegetti, S., McKean, J. P., Spingola, C., Stacey, H. R., & Fassnacht, C. D. 2022, A lensed radio jet at milliarcsecond resolution I: Bayesian comparison of parametric lens models. *MNRAS*, 516(2), 1808–1828.

Springel, V., Wang, J., Vogelsberger, M., Ludlow, A., Jenkins, A., Helmi, A., Navarro, J. F., Frenk, C. S., & White, S. D. M. 2008, The Aquarius Project: the subhaloes of galactic haloes. *MNRAS*, 391(4), 1685–1711

Van de Vyvere, L., Gomer, M. R., Sluse, D., Xu, D., Birrer, S., Galan, A., & Vernardos, G. 2022, TDCOSMO. VII. Boxyness/discyness in lensing galaxies: Detectability and impact on H_0. *A&A*, 659, A127.

Vegetti, S., Birrer, S., Despali, G., Fassnacht, C. D., Gilman, D., Hezaveh, Y., Perreault Levasseur, L., McKean, J. P., Powell, D. M., O'Riordan, C. M., & Vernardos, G. 2023, Strong gravitational lensing as a probe of dark matter. *arXiv e-prints*, arXiv:2306.11781.

Vegetti, S. & Koopmans, L. V. E. 2009, Bayesian strong gravitational-lens modelling on adaptive grids: objective detection of mass substructure in Galaxies. *MNRAS*, 392(3), 945–963.

Vegetti, S., Koopmans, L. V. E., Auger, M. W., Treu, T., & Bolton, A. S. 2014, Inference of the cold dark matter substructure mass function at z = 0.2 using strong gravitational lenses. *MNRAS*, 442(3), 2017–2035.

Strong Gravitational Lensing in the Era of Big Data
Proceedings IAU Symposium No. 381, 2024
H. Stacey, A. Sonnenfeld & C. Grillo, eds.
doi:10.1017/S1743921323004866

Effectively Investigating Dark Matter Microphysics With Strong Gravitational Lensing Anisotropies

Birendra Dhanasingham

Department of Physics and Astronomy, University of New Mexico, 210 Yale Blvd NE, Albuquerque, NM 87106, USA. email: birendradh@unm.edu

Abstract. Over the last two decades, strong gravitational lensing has emerged as a potential method for studying the nature and distribution of dark matter on sub-galactic scales. In addition to the main lens substructure, line-of-sight dark matter haloes contribute greatly to the subtle perturbations of lensed images. Line-of-sight haloes, unlike dark matter subhaloes, imprint distinct anisotropic and quadrupole signatures in the maps that depict the divergence and curl of the effective deflection field, respectively, giving rise to quadrupole moments of the image-plane averaged two-point correlation function of these maps. In terms of central density evolution and dark matter halo distribution, the shapes and amplitudes of the two-point function multipoles alter dramatically in the presence of warm dark matter and self-interacting dark matter. This method, in conjunction with upcoming large-scale surveys, provides the prospect of improving the constraints on dark matter at a critical time in strong gravitational lensing research.

Keywords. gravitational lensing: strong – galaxies: haloes – methods: numerical – dark matter

1. Introduction

Dark matter plays a critical role in the hierarchical structure formation in the classic cold dark matter (CDM) model, where dark matter is spread into haloes with a wide range of masses. While the largest haloes are known to contain clusters and galaxies and hence are luminous, the smallest haloes have a relatively Weak gravitational potential to hold gases and thus are dark. The investigation of such dark haloes, particularly on small distance scales, might give answers to the inadequacy of the Lambda-CDM model's success deep in the non-linear regime.

Strong gravitational lensing is a unique method for studying dark matter haloes at sub-galactic distance scales. Dhanasingham *et al.* (2023a) pointed out that line-of-sight dark matter haloes located between the lensed source and the observer produce a distinct anisotropic signature in the two-point correlation function of the lensing deflection field when considering the collective effect of multiplane gravitational lens planes. I summarize here this anisotropic correlation function approach presented by Dhanasingham *et al.* (2023a) as a potential probe of dark matter, as well as how this function behaves in the presence of different dark matter physics, as discussed in Dhanasingham *et al.* (2023b).

2. Effective Multiplane Lensing

Photons are deflected repeatedly by dark matter haloes as they travel across the Universe from a background source to an observer. This multiplane lensing is often handled using a recursive technique that involves solving a lens equation for each lens plane

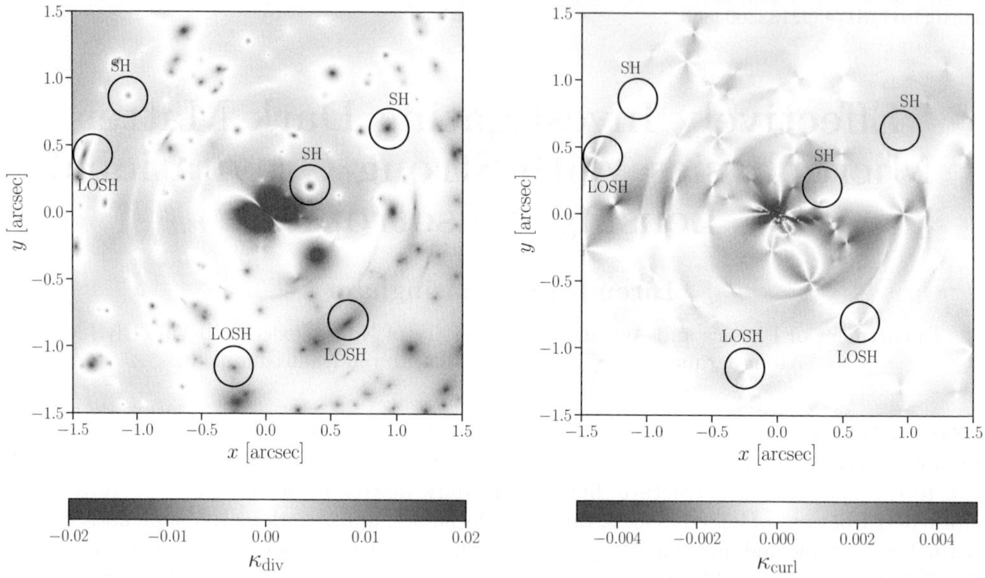

Figure 1. Divergence (left) and curl (right) of the effective deflection field for a CDM model, incorporating contributions from subhaloes (SH) and line-of-sight haloes (LOSH). On the left, line-of-sight haloes seem deformed in the angular direction, although subhaloes are mainly circular, and on the right, these line-of-sight haloes are accompanied by quadrupolar structures, whereas subhaloes contribute nothing to this map. Figure from Dhanasingham *et al.* (2023b).

(Blandford & Narayan 1986; McCully *et al.* 2014). Lensing, on the other hand, is intrinsically a two-dimensional map from the source plane to the image plane. This map may be written as a pure gradient of some scalar lensing potential in the single-plane case. Because of the nonlinear coupling between the multiple lens planes, this map is no longer a pure gradient for multiplane lensing. Dhanasingham *et al.* (2023a) thus proposed a novel approach named "effective multiplane lensing" that took into account the collective influence of all the lens planes in the strong lens system, where one scalar ($\phi_{\rm eff}$) and one vector ($\mathbf{A}_{\rm eff}$) effective potential were included to write down the effective deflection field

$$\boldsymbol{\alpha}_{\rm eff}(\mathbf{x}) = \boldsymbol{\nabla}\phi_{\rm eff}(\mathbf{x}) + \boldsymbol{\nabla} \times \mathbf{A}_{\rm eff}(\mathbf{x}), \tag{1}$$

and thus the lens equation

$$\mathbf{u}(\mathbf{x}) = \mathbf{x} - \boldsymbol{\alpha}_{\rm eff}(\mathbf{x}), \tag{2}$$

where $\mathbf{u}$ and $\mathbf{x}$ are the source plane and image plane coordinate systems, respectively. This decomposition of the effective deflection field is based on Helmholtz's theorem. Without any approximation, the two effective scalar and vector potentials encode the full nonlinear multiplane lensing map from the source plane to the image plane. In this terminology, the two forms of two-dimensional projected mass densities are defined as $\kappa_{\rm div} \equiv \frac{1}{2}\boldsymbol{\nabla}\cdot\boldsymbol{\alpha}_{\rm eff} - \kappa_0$ and $\kappa_{\rm curl} \equiv \frac{1}{2}\boldsymbol{\nabla}\times\boldsymbol{\alpha}_{\rm eff}\cdot\hat{\mathbf{z}}$, taking into account the divergence and curl of the effective deflection field (Gilman *et al.* 2019; Çağan Şengül *et al.* 2020; Dhanasingham *et al.* 2023a). To distinguish the collective effect from dark matter haloes, the main lens convergence κ_0 is deducted from the $\kappa_{\rm div}$ equation. Figure 1 depicts the $\kappa_{\rm div}$ and $\kappa_{\rm curl}$ for a CDM model.

As seen in the left panel of Figure 1, Line-of-sight haloes appear stretched in the tangential direction and generate arc-like patterns due to the distortion on their deflection field caused by the main lens due to the non-linear nature of multiplane gravitational lensing. While these distortions contribute anisotropic signals to the $\kappa_{\rm div}$ map, subhaloes still seem circular and hence stay isotropic in the $\kappa_{\rm div}$ map. When comparing the $\kappa_{\rm div}$ and $\kappa_{\rm curl}$ maps, one can see quadrupolar patterns in the right panel at the sites of line-of-sight haloes. These anisotropic and quadrupolar fingerprints, as proposed by Dhanasingham *et al.* (2023a), can be utilized to identify line-of-sight dark matter haloes from the main lens substructure and study dark matter microphysics of these haloes. Dhanasingham *et al.* (2023a,b) proposed a method for capturing the useful information encoded in these signatures based on the two-point correlation function and its breakdown into multipoles, as described in the next section.

2.1. *Two-point Correlation Functions and Multipoles*

As previously noted, the existence of line-of-sight haloes causes a noticeable anisotropy between the tangential and radial directions in the divergence of the effective deflection field and the quadrupolar patterns in the curl of the deflection field. Because of these anisotropies and quadrupoles, the image-plane averaged two-point functions of the $\kappa_{\rm div}$ and $\kappa_{\rm curl}$ fields have non-vanishing quadrupole moments (Dhanasingham *et al.* 2023a,b). The two-point function for a given κ (represents either $\kappa_{\rm div}$ or $\kappa_{\rm curl}$) is denoted as:

$$\xi(\mathbf{r}) = \frac{1}{A}\int_A \mathrm{d}^2\mathbf{r}_1 \,\left[\kappa(\mathbf{r}_1) - \langle\kappa(\mathbf{r}_1)\rangle\right]\,\left[\kappa(\mathbf{r}_2) - \langle\kappa(\mathbf{r}_2)\rangle\right], \tag{3}$$

where $\mathbf{r}_1$ and $\mathbf{r}_2 = \mathbf{r}_1 + \mathbf{r}$ are the position vectors of two points on the map connected by vector $\mathbf{r}$, and the A is the area of the image where the correlation is computed. The multipole decompositions for the two-point functions of the $\kappa_{\rm div}$ and $\kappa_{\rm curl}$ maps are given by

$$\xi_{{\rm div},\ell}(r) = \frac{2-\delta_{\ell 0}}{\pi}\int_0^{\pi} \mathrm{d}\theta\; \xi_{\rm div}(r,\theta)\cos(\ell\theta) \tag{4}$$

and

$$\xi_{{\rm curl},\ell}(r) = \frac{2}{\pi}\int_0^{\pi} \mathrm{d}\theta\; \xi_{\rm curl}(r,\theta) A_\theta \sin(\ell\theta), \tag{5}$$

respectively. Here θ is the angle between $\mathbf{r}$ and radial direction and δ_{ij} is the Kronecker delta. The parameter $A_\theta = \begin{cases} +1 & 0 \le \theta < \frac{\pi}{2}, \\ -1 & \frac{\pi}{2} \le \theta < \pi \end{cases}$ in equation (5) captures the useful information hidden in the odd-parity structure in the $\kappa_{\rm curl}$ map, as detailed in Dhanasingham *et al.* (2023b).

Figure 2 depicts the multipole decompositions of the two-point correlation function of the projected mass densities $\kappa_{\rm div}$ and $\kappa_{\rm curl}$ for a population of CDM subhaloes and line-of-sight haloes following this approach. The statistical features of the main-lens subhaloes are completely represented by a monopole ($\ell = 0$) term (showing no anisotropy), but the line-of-sight haloes contribute to both the monopole and the quadrupole ($\ell = 2$) moments as shown in the left panel. This quadrupole moment represents the anisotropies in the deflection field perturbation induced by line-of-sight haloes between the tangential and radial directions. Using Fisher forecasting, Dhanasingham *et al.* (2023a) pointed out that present space-based telescopes and future extremely large telescopes may identify the quadrupole signal. Because the main lens substructure makes no contribution to the $\kappa_{\rm curl}$ map, the non-zero quadrupole moment presented in the right panel solely reflects the statistical features of the line-of-sight haloes (Dhanasingham *et al.* 2023b).

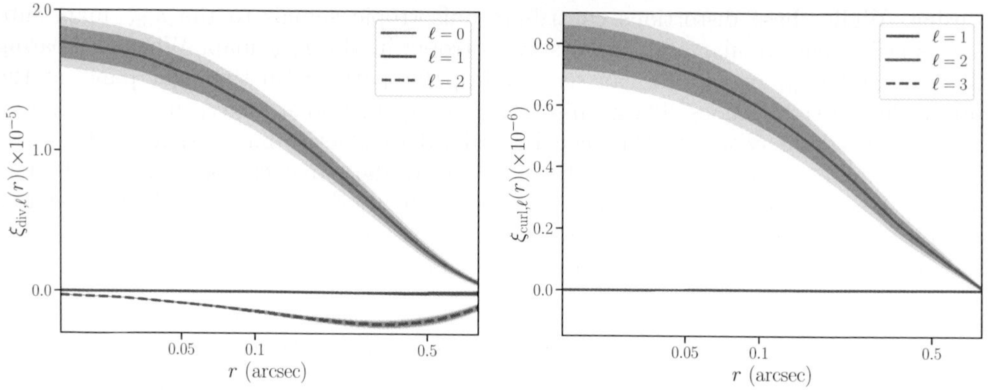

Figure 2. Multipole moments of the two-point correlation function of a CDM model's κ_{div} (left) and κ_{curl} (right) fields. On the left panel, the monopole is formed by both the line-of-sight haloes and the primary lens substructure, but the non-zero quadrupole is formed only by the line-of-sight haloes. The non-zero quadrupole moment in the right panel is entirely due to the line-of-sight haloes.

3. Multipoles and Dark Matter Microphysics

Dhanasingham *et al.* (2023b) investigated the use of two-point function multipoles as a probe of dark matter from two perspectives: first, they investigated how the shapes and amplitudes of multipoles change with the evolution of the central density in self-interacting dark matter (SIDM) haloes, and second, they investigated how the abundance of dark matter haloes affects the multipoles in warm dark matter models. I describe their findings in this section.

3.1. *Self-interacting Dark Matter (SIDM)*

By enabling dark matter particles to interchange energy and momentum, self-interactions can radically modify the internal dynamics of haloes. Initially, a Navarro-Frenk-White (NFW) (Navarro *et al.* 1996) halo has a high temperature close to the outer skirts, so heat flows inward and particles flow outward due to dark matter particle self-interaction until a low-density isothermal core forms, and this core, with its high temperature relative to the outer skirts, now favors an outward heat flow. The isothermal core eventually begins to shrink, resulting in a sharp density cusp. When compared to cored haloes, these core-collapsed haloes serve as excellent lenses. The number of core-collapsed haloes grows as the self-interaction cross-section increases, but the central densities of cored haloes drop due to the high rate of outward mass flow.

The correlation function multipoles of the κ_{div} and κ_{curl} maps for SIDM models are shown in the left panel of Figure 3 as a function of the self-interaction cross-section, $\sigma(\propto \sigma_0)$. When σ is small, cored haloes dominate the amplitudes of the multipoles of the two-point functions, resulting in a lower relative amplitude than the CDM model. As σ increases, subhaloes can begin to core-collapse, resulting in a steeper and stronger monopole ($\xi_{\mathrm{div},0}$) for lower r values, but line-of-sight haloes are still cored, resulting in comparatively suppressed quadrupole ($\xi_{\mathrm{div},2}$) compared to the CDM model. Subhaloes and line-of-sight haloes both experience core-collapse as σ further increases, causing $\xi_{\mathrm{div},0}$ and $\xi_{\mathrm{div},2}$ to grow large and steep at small r values. In addition, due to the core-collapsed line-of-sight haloes, quadrupole, $\xi_{\mathrm{curl},2}$, exhibits a slight amplitude increase at small r. As a result, it is obvious that the monopole and quadrupole moments of the correlation function multipoles could be used to investigate dark matter self-interactions.

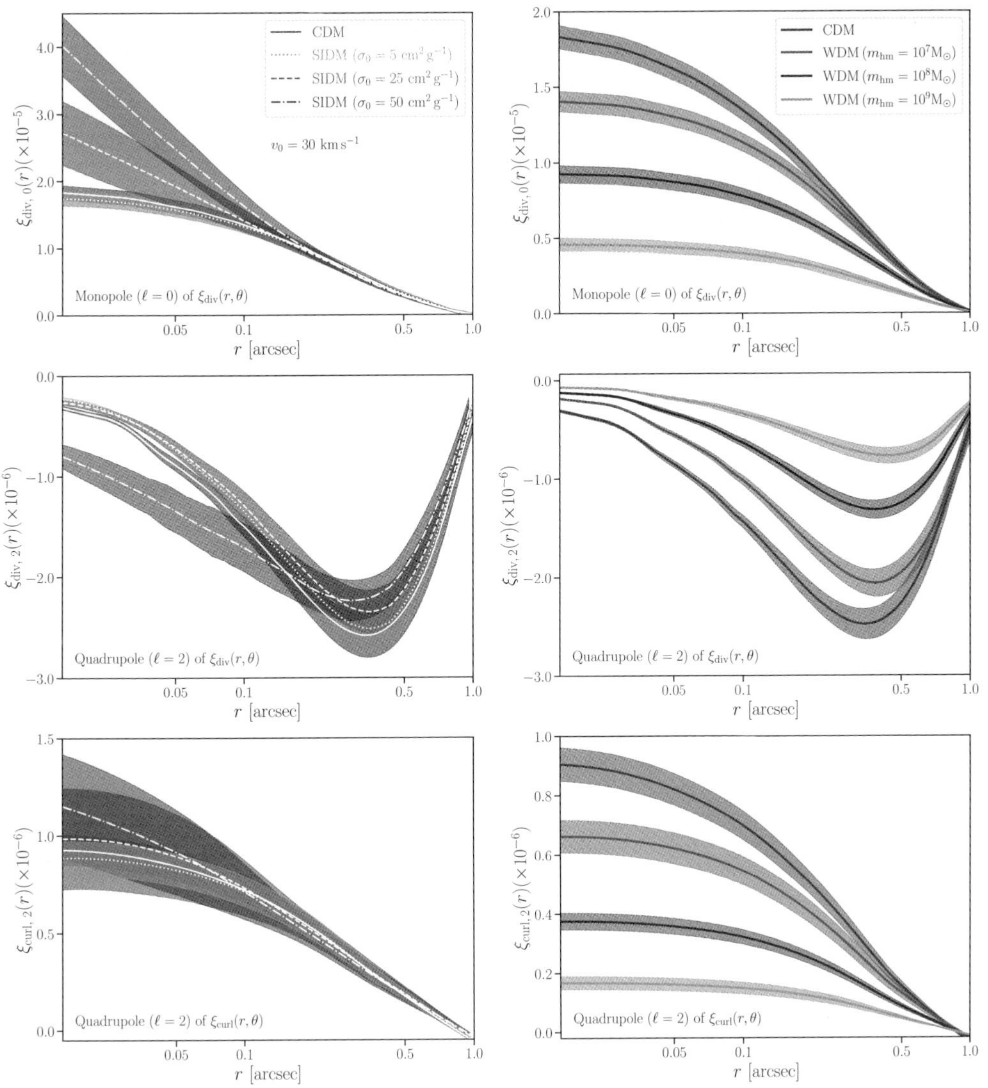

Figure 3. The monopole (top panel) and the quadrupole (middle panel) of the two-point correlation function $\xi_{\rm div}$, and the quadrupole (bottom panel) of the $\xi_{\rm curl}$ function for SIDM (left panel) and WDM (right panel) models. The coloured regions represent the 68% credible intervals. The amplitude and shapes of the multipoles are principally determined by the evolution of the central density in SIDM models and the abundance in WDM models. Figure from Dhanasingham *et al.* (2023b).

3.2. *Warm Dark Matter (WDM)*

The right panel of Figure 3 depicts the correlation function multipoles of the $\kappa_{\rm div}$ and $\kappa_{\rm curl}$ maps for three WDM models. As the characteristic half-mode mass ($m_{\rm hm}$) grows in WDM models, the abundance of dark matter haloes diminishes because free-streaming effects erase the structure below $m_{\rm hm}$. The relative amplitudes of the multipoles fall as $m_{\rm hm}$ increases at small radial distances, reflecting the suppressed abundance of small haloes. The difference in amplitudes is small over large radial distances because the massive haloes that dominate these scales are less impacted by free-streaming effects.

As a result, the combination of quadrupole and monopole moments includes crucial information on the physics beneath the minuscule dark matter haloes. Observing it might put our traditional theory of structure formation to the test.

References

Blandford, R. & Narayan, R. 1986, *ApJ*, 310, 568

Çağan Şengül, A., Tsang, A., Diaz Rivero, A., Dvorkin, C., Zhu, H. & Seljak, U. 2020, *Phys. Rev. D*, 102, e063502

Dhanasingham, B., Cyr-Racine, F., Peter, A., Benson, A. & Gilman, D. 2023, *MNRAS*, 518, 5843

Dhanasingham, B., Cyr-Racine, F., Mace, C., Peter, A. & Benson, A. 2023, *MNRAS*, 526, 5455

Gilman, D., Birrer, S., Treu, T., Nierenberg, A. & Benson, A. 2019, *MNRAS*, 487, 5721

McCully, C., Keeton, C., Wong, K. & Zabludoff, A. 2014, *MNRAS*, 443, 3631

Navarro, J., Frenk, C. & White, S. 1996, *ApJ*, 462, 563

Chapter 3

Strong Gravitational Lensing in the Era of Big Data
Proceedings IAU Symposium No. 381, 2024
H. Stacey, A. Sonnenfeld & C. Grillo, eds.
doi:10.1017/S1743921323004076

High-precision strong lensing models of galaxy clusters in the JWST era

P. Bergamini[1,2], C. Grillo[1,5], P. Rosati[2,3,9], A. Acebron[1,5], E. Vanzella[2], A. Mercurio[4,10], M. Meneghetti[2], G. Angora[3,4], G. Granata[1,5], U. Mestric[1,2], G. B. Caminha[6,7] and T. Treu[11]

[1]Dipartimento di Fisica, Università di Milano, via Celoria 16, I-20133 Milano, Italy.
email: pietro.bergamini@unimi.it

[2]INAF – OAS, Osservatorio di Astrofisica e Scienza dello Spazio di Bologna, via Gobetti 93/3, I-40129 Bologna, Italy

[3]Dipartimento di Fisica e Scienze della Terra, Università di Ferrara, Via Saragat 1, I-44122 Ferrara, Italy

[4]INAF – Osservatorio Astronomico di Capodimonte, Salita Moiariello 16, I-80131 Napoli, Italy

[5]INAF – IASF Milano, via A. Corti 12, I-20133 Milano, Italy

[6]Technische Universität München, Physik-Department, James-Franck Str. 1, D-85741 Garching, Germany

[7]Max-Planck-Institut für Astrophysik, Karl-Schwarzschild-Str. 1, D-85748 Garching, Germany

[8]INAF – Osservatorio Astronomico di Trieste, via G. B. Tiepolo 11, I-34131 Trieste, Italy

[9]INFN, Sezione di Ferrara, Via Saragat 1, I-44122 Ferrara, Italy

[10]Dipartimento di Fisica, Università di Salerno, Via Giovanni Paolo II, 132, I-84084, Fisciano (SA), Italy

[11]Department of Physics and Astronomy, University of California, Los Angeles, 430 Portola Plaza, Los Angeles, CA 90095, USA

Abstract. We present high-precision strong lensing models for the galaxy clusters MACS J0416.1−0403 at $z = 0.396$ and Abell 2744 at $z = 0.307$. The models are constrained by two of the largest data-sets of secure multiple images ever used in lensing. These are identified from the photometric images observed by the *Hubble* space telescope and *JWST* in combination with spectroscopic data obtained by the Multi-Unit Spectroscopic Explorer at the Very Large Telescope. The same spectro-photometric data are used to create pure and complete samples of cluster member galaxies. Our models allow an extremely precise estimation of the cluster total mass distribution and produce accurate magnification maps that are fundamental to study the physical properties (mass, size, luminosity, etc.) of the lensed high-redshift galaxies.

Keywords. Galaxy Cluster, Strong Gravitational Lensing, Dark Matter

1. Introduction

The currently accepted cosmological model (a.k.a. ΛCDM) assumes that dark energy (DE) and dark matter (DM) are the two dominant components of the Universe, accounting for more than 95% of its total mass-energy budget, while baryonic matter contributes just for ∼5%. While DE is responsible for the measured accelerated expansion of the Universe, DM is expected to be formed by non-relativistic (cold) and weakly interacting particles whose nature is still unknown. Despite numerous particle physics experiments were designed to detect the DM particles directly, the only observation evidence of the DM existence remains its gravitational interaction with baryonic matter in astronomical systems.

In this context, the galaxy clusters, which are the largest gravitationally bound structures in the Universe and with the $\sim 85\%$ of their total mass made of DM, represent unique nature laboratories to study the DM particles. In particular, a detailed analysis of the total, baryonic, and DM mass density distributions in the inner region of galaxy clusters can provide stringent constraints on the DM physical properties. In fact, both the slope of the radial profile of the DM in the central region of galaxy clusters and the spatial segregation of the DM subhalos with respect to the baryonic matter are strongly affected by the interaction cross-section of the DM particles. Over the last decade, strong gravitational lensing (SL) has become the most accurate technique to determine the projected total mass distribution of the core region of galaxy clusters. The reasons for its success have to be mainly ascribed to the high-quality spectro-photometric data obtained by the *Hubble* Space Telescope (HST), and by the VIMOS (VIsible MultiObject Spectrograph, Le Fèvre et al. 2003) and MUSE (Multi-Unit Spectroscopic Explorer, Bacon et al. 2012) spectrographs mounted on the Very Large Telescope (VLT). These data are used to identify large samples of spectroscopically confirmed multiple images (e.g., Grillo et al. 2016; Karman et al. 2017; Lagattuta et al. 2017; Richard et al. 2021; Bergamini et al. 2021, 2023; Caminha et al. 2016, 2019) and pure and complete catalogs of cluster member galaxies (e.g., Bergamini et al. 2019; Angora et al. 2020). With the advent of the JWST, a further step forward was possible in the SL modeling of galaxy clusters thanks to the photometric and spectroscopic data of unprecedented quality.

Other than to study the mass distribution of galaxy clusters, the SL can be used to exploit the clusters as cosmic telescoped to study a population of background lensed sources that are the progenitor of the present-day galaxies and globular clusters, and that may have played a fundamental role in the re-ionization of the Universe (e.g., Vanzella et al. 2017, 2020, 2022, 2023; Meštrić et al. 2022; Roberts-Borsani et al. 2023). In particular, the magnification maps obtained by the SL models are fundamental to derive the intrinsic physical proprieties of these fain, highly magnified, and high-z sources.

In this contribution, we will present state-of-the-art strong lensing models for the two massive galaxy clusters MACS J0416.1−0403 (hereafter M0416) at $z = 0.396$ and Abell 2744 (hereafter A2744) at $z = 0.307$. Throughout this work, we assume a flat ΛCDM cosmology with $\Omega_{\rm m} = 0.3$ and $H_0 = 70\,{\rm km\,s^{-1}\,Mpc^{-1}}$.

2. Observational data

The galaxy cluster M0416 was one of the selected targets of the Cluster Lensing And Supernova survey with Hubble (CLASH, Postman et al. 2012) and Hubble Frontier Fields (HFF, Lotz et al. 2017) program that provided multi-band HST observations of the cluster. Several spectroscopic campaigns were also carried out on the cluster to complement the photometric data. In particular, the CLASH-VLT (P.I.: Rosati) program provided VLT/VIMOS observations that allowed the measurement of the redshift of thousands of sources over a 20 arcmin^2 field-of-view. The following MUSE pointings were also obtained on the cluster: a MUSE pointing of 2h centered on the northeast (NE) region of the cluster (GTO 094.A-0115B, P.I.: Richard); a MUSE pointing of 11h centered on the southwest (SW) region of the cluster (094.A0525(A), P.I.: Bauer); an ultra-deep MUSE pointing centered on the northeast region of the cluster that, when stacked with previous observations, allows to reach a total exposure time of 17.1h in the NE region (0100.A-0763(A), P.I.: Vanzella, Vanzella et al. 2021).

Similarly, the HFF program provided deep HST photometric data on the core region of the galaxy cluster A2744, while shallower HST observations on the external regions were obtained during the Beyond Ultra-deep Frontier Fields And Legacy Observations (BUFFALO, Steinhardt et al. 2020) campaign. More recently the cluster was selected as one the the primary targets of the JWST. In particular, NIRCam images were taken within the GLASS-JWST program ERS-1324 (P.I.: Treu, Treu et al. 2022), the Ultradeep NIRSpec and NIRCam ObserVations before the Epoch of Reionization Cycle 1 Treasury program (UNCOVER, GO-2561, co-P.I.s: Labbè and Bezanson, Bezanson et al. 2022), and the DDT program 2756 (P.I.: Chen). These data were complemented by a total of 10 MUSE pointings of 1 arcmin^2 each. Five of the pointings are centered in the core region of the cluster and have an exposure time between 2h and 5h (Richard et al. 2021; Mahler et al. 2018), while the other five (co-P.I.s: Mason, Vanzella) have an exposure of 1h each and mostly cover the same area of the GLASS-JWST ERS NIRCam observations.

3. Lens modeling technique

To develop the strong lensing models we exploit the publicly available parametric software *LensTool* (Kneib et al. 1996; Jullo et al. 2007; Jullo & Kneib 2009). In parametric software, the total mass distribution of the cluster has to be decomposed into a sum of different parametric components:

$$\phi_{tot} = \sum_{i=1}^{N_h} \phi_i^{halo} + \sum_{j=1}^{N_{gas}} \phi_j^{gas} + \sum_{k=1}^{N_g} \phi_k^{gal} + \phi_{foreg}. \tag{1}$$

The first term of the sum is used to parameterize the large cluster-scale halos of the galaxy cluster that are primarily made of DM. The second term describes the hot gas (baryonic) mass contained in the cluster. The third term accounts for the contribution of the cluster galaxies to the cluster total mass. These are usually described with truncated Singular Isothermal Sphere (tSIS) profiles whose free parameters, i.e., their central velocity dispersion (σ_i) and truncation radius ($r_{cut,i}$), are scaled with the galaxy luminosity (L_i) according to the following two scaling relations:

$$\sigma_i = \sigma^{ref} \left(\frac{L_i}{L_{ref}} \right)^{\alpha}, \quad r_{cut,i} = r_{cut}^{ref} \left(\frac{L_i}{L_{ref}} \right)^{\beta_{cut}}, \tag{2}$$

where the two normalizations σ^{ref} and r_{cut}^{ref}, and the two slopes α and β_{cut} are free parameters of the lens model. Finally, the fourth term in Eq. 1 is necessary to consider the non-negligible impact that foreground galaxies and/or massive structures residing in the outer region of the cluster have on the lensing observables.

The lens model optimization is performed using a Bayesian approach by looking for the combination of model free parameters, $\boldsymbol{\xi}$, that minimize the following χ^2 function:

$$\chi^2(\boldsymbol{\xi}) := \sum_{j=1}^{N_{fam}} \sum_{i=1}^{N_{im}^j} \left(\frac{\left\| \mathbf{x}_{i,j}^{pred}(\boldsymbol{\xi}) - \mathbf{x}_{i,j}^{obs} \right\|}{\Delta x_{i,j}} \right)^2, \tag{3}$$

where $\mathbf{x}_{i,j}^{obs}$ and $\mathbf{x}_{i,j}^{pred}(\boldsymbol{\xi})$ are the observed and model predicted positions of the i-th multiple image of the j-th background source (N_{fam} is the total number of background sources), respectively.

4. Description of the galaxy cluster strong lensing models

In this section, we briefly describe the lens models for the two galaxy clusters M0416 and A2744. We refer to the papers by Bergamini et al. (2023b) and Bergamini et al. (2023c) for a detailed description of the models.

4.1. *M0416 lens model*

The cluster-scale component of M0416 (ϕ^{halo} in Eq. 1) is described as a combination of four non-truncated dual pseudo-isothermal elliptical mass distributions (dPIEs, Limousin et al. 2005; Elíasdóttir et al. 2007; Bergamini et al. 2019). Two elliptical dPIEs have centers close to the position of the two brightest cluster galaxies (BCGs), while a third elliptical dPIE is used to provide second-order corrections to the total mass in the SW region of the cluster. Finally, the fourth circular dPIE is used to account for an over-density of galaxies located NE of the northern BCG. On the whole, the cluster-scale component gives 22 free parameters to the lens model.

For the hot gas component (ϕ^{gas} in Eq. 1), we assume the mass parameterization obtained by Bonamigo et al. (2018) from an analysis of the *Chandra* X-ray observations. The latter is described as the sum of four dPIEs and does not add any free parameter to the lens model.

The subhalo mass component of the cluster (ϕ^{gal} in Eq. 1) consists of 212 galaxies parameterized through the scaling relations in Eq. 2. Following the analysis by Bergamini et al. (2019), we infer the values of $\alpha = 0.30$ and $\beta_{cut} = 0.60$ from the measured stellar kinematics of 64 member galaxies (down to $mag_{\mathrm{F160W}} \sim 22$) that is obtained by fitting the high-quality MUSE galaxy spectra with the software *pPXF* (see Cappellari & Emsellem 2004; Cappellari 2017, 2023). The same measurements are also used to define a Gaussian prior on the normalization σ^{ref}. An additional core-less dPIE is used to describe the cluster galaxy Gal-8971 (as detailed in Bergamini et al. 2023b) forming a galaxy-scale strong lensing system. The subhalo mass component adds six more free parameters to the lens model.

Finally, a tSIS profile is used to account for the impact on the lens model coming from a massive foreground galaxy residing in the SW region of the cluster. The galaxy is at a redshift $z = 0.112$. Thus, the cluster lens model is characterized by 30 free parameters that have to be inferred from the observed position of the multiple images.

A careful inspection of the HST photometric and MUSE spectroscopic data allowed us to identify a total of 237 spectroscopically confirmed multiple images from 88 background sources with a redshift range $0.94 \leq z \leq 6.63$. Among the multiple images, we include several multiply imaged substructures of lensed extended sources since these have proven to be extremely useful in constraining the local magnification and the position of the critical lines. We note that this is the largest dataset of secure multiple images ever used in strong gravitational lensing. The observed multiple images provide 298 constraints to the lens model, such that the number of model degree of freedom (dof) is 268.

4.2. *A2744 lens model*

The cluster-scale component of the A2744 lens model is described by using four dPIEs; two are centered on the two main cluster BCGs, one is located between the two bright galaxies G1 and G2 (see Figure 1), and one is centered on the external galaxy G3. Thus, the cluster-scale component counts a total of 22 free parameters.

The subhalo mass component of the cluster contains a sample of 177 cluster member galaxies, five of which (i.e., the two BCGs and the external galaxies G1, G2, and G3) are modeled as core-less dPIEs (a tSIS is used for G3) and 172 are described through the

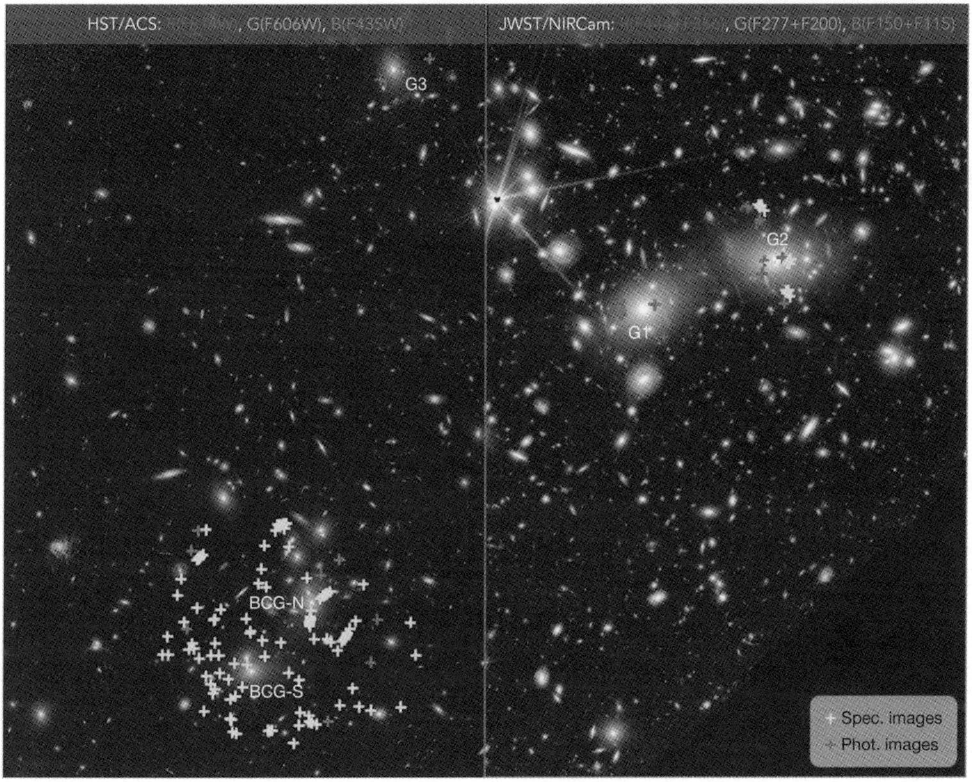

Figure 1. Color-composite merging images of the galaxy cluster A2744. The left half of the image is obtained by a combination of the F814W, F606W, and F435W HST/ACS filters, while the right half of the image is obtained by combining the JWST/NIRCam F444, F356, F277, F200, F150, and F115 filters. The 121 spectroscopically confirmed and the 28 photometric multiple images used in the lens model are marked with green and red crosses, respectively.

scaling relations in Eq. 2. The measured stellar kinematics of 85 member galaxies (down to $mag_{\mathrm{F160W}} \sim 22$) is used to determine the slope values $\alpha = 0.40$ and $\beta_{cut} = 0.41$, but none Gaussian prior is assumed on the normalization σ^{ref}.

Thanks to the newly observed JWST/NIRCam images of the cluster and a careful inspection of HST and MUSE data, we were able to identify 149 multiple images from 50 background sources. 121 images are spectroscopically confirmed with a redshift range $1.03 \leq z \leq 9.76$. The redshifts of the remaining 28 photometric images, from 8 background sources, are additional free parameters of the lens model. By considering these 8 additional unknowns, the total number of model free parameter is thus equal to 50, while the multiple images provide 198 constraints to the model (dof=148).

5. Discussion and conclusions

We have presented two high-precision strong lensing models for the galaxy clusters M0416 ($z = 0.396$) and A2744 ($z = 0.307$) based on deep photometric images of the HST and JWST (for A2744) complemented with high-resolution spectroscopic data obtained with MUSE. The lens models are constrained by two of the largest samples of spectroscopically confirmed multiple images ever used in strong gravitational lensing, counting 237 and 149 images for M0416 and A2744, respectively. Both the models have a precision of just $\Delta_{rms} = 0.43$" in reproducing the position of the multiple images (Bergamini et al. 2023b,c).

In Bergamini et al. (2023b), we presented a comparison between the M0416 lens model described in this work with other published models for the same cluster and exploiting similar data-sets. The comparison is performed by exploiting the *GravityFM* forward modeling code (Bergamini et al. in prep.) to test the capabilities that the different models have in reproducing the observed shapes and magnifications of the surface brightness of several extended lensed galaxies. This analysis demonstrates that our M0416 lens model outperforms all the others in producing accurate deflection and magnification maps even on the scale of the galaxies (Vanzella et al. 2023; Meštrić et al. 2022). Moreover, the analyses presented in Bergamini et al. (2023b) demonstrate that our model is able to robustly characterize the M0416 total mass distribution.

The large spatial coverage and resolution of the JWST/NIRCam images of A2744 allowed us to identify a large sample of multiple images (149 in total) distributed over an area of ~ 30 arcmin^2. Remarkably, the photometric redshifts measured for 6 out of 8 of the non-spectroscopic lensed sources included in the model are strongly consistent with the model-predicted redshifts. In the last year, our A2744 lens model was adopted as the reference model by the GLASS-JWST collaboration and was widely used to predict the magnification factors that are necessary to study the physical properties (e.g., star-formation rates, stellar masses, sizes) of the lensed high-z galaxies (Roberts-Borsani et al. 2023; Castellano et al. 2023).

The M0416 and A2744 lens models presented in this work are publicly available at the following link: `www.fe.infn.it/astro/lensing/`. All model inputs and outputs are accessible through a newly developed Strong Lensing Online Tool (SLOT). This is a user-friendly graphical interface that allows researchers to take full advantage of the statistical results and predictive power of our lens models.

6. Future perspectives

The presented analysis and lens modeling technique will play a fundamental role in the near future when a large number of galaxy cluster lenses are expected to be observed by the JWST and by large area surveys such as the Large Synoptic Survey Telescope (LSST) and *Euclid*. In particular, by combining the magnification power of the cluster lenses with the new generation telescopes (e.g., JWST and ELT) it will be possible to study the physical properties of some of the first sources ensembled in the Universe.

Acknowledgements

Support for program JWST-ERS-1324 was provided by NASA through a grant from the Space Telescope Science Institute, which is operated by the Association of Universities for Research in Astronomy, Inc., under NASA contract NAS 5-03127. The Hubble Frontier Field program (HFF) and the Beyond Ultra-deep Frontier Fields And Legacy Observations (BUFFALO) are based on the data made with the NASA/ESA *Hubble Space Telescope*. The Space Telescope Science Institute is operated by the Association of Universities for Research in Astronomy, Inc., under NASA contract NAS 5-26555. ACS was developed under NASA Contract NAS 5-32864. Based also on observations collected at the European Southern Observatory for Astronomical research in the Southern Hemisphere under ESO programmes with IDs 094.A-0115 (PI: Richard). We acknowledge financial support through grants PRIN-MIUR 2015W7KAWC, 2017WSCC32, and 2020SKSTHZ. AA has received funding from the European Union's Horizon 2020 research and innovation programme under the Marie Skłodowska-Curie grant agreement No 101024195 - ROSEAU. GBC thanks the Max Planck Society for support through the Max Planck Research Group for S. H. Suyu and the academic support from the German Centre for Cosmological Lensing. MM acknowledges support from the Italian Space Agency (ASI) through contract "Euclid - Phase D".

References

Angora, G., Rosati, P., Brescia, M., et al. 2020, A&A, 643, A177. doi:10.1051/0004-6361/202039083

Bacon, R., Accardo, M., Adjali, L., et al. 2012, The Messenger, 147, 4

Bergamini, P., Acebron, A., Grillo, C., et al. 2023, ApJ, 952, 84. doi:10.3847/1538-4357/acd643

Bergamini, P., Grillo, C., Rosati, P., et al. 2023, A&A, 674, A79. doi:10.1051/0004-6361/202244834

Bergamini, P., Rosati, P., Mercurio, A., et al. 2019, A&A, 631, A130. doi:10.1051/0004-6361/201935974

Bergamini, P., Rosati, P., Vanzella, E., et al. 2021, A&A, 645, A140. doi:10.1051/0004-6361/202039564

Bergamini, P., Acebron, A., Grillo, C., et al. 2023, A&A, 670, A60. doi:10.1051/0004-6361/202244575

Bezanson, R., Labbe, I., Whitaker, K. E., et al. 2022, arXiv:2212.04026. doi:10.48550/arXiv.2212.04026

Bonamigo, M., Grillo, C., Ettori, S., et al. 2018, ApJ, 864, 98. doi:10.3847/1538-4357/aad4a7

Caminha, G. B., Grillo, C., Rosati, P., et al. 2016, A&A, 587, A80. doi:10.1051/0004-6361/201527670

Caminha, G. B., Rosati, P., Grillo, C., et al. 2019, A&A, 632, A36. doi:10.1051/0004-6361/201935454

Cappellari, M. & Emsellem, E. 2004, PASP, 116, 138. doi:10.1086/381875

Cappellari, M. 2017, MNRAS, 466, 798. doi:10.1093/mnras/stw3020

Cappellari, M. 2023, MNRAS. doi:10.1093/mnras/stad2597

Castellano, M., Fontana, A., Treu, T., et al. 2023, ApJ, 948, L14. doi:10.3847/2041-8213/accea5

Elíasdóttir, Á., Limousin, M., Richard, J., et al. 2007, arXiv:0710.5636. doi:10.48550/arXiv.0710.5636

Grillo, C., Karman, W., Suyu, S. H., et al. 2016, ApJ, 822, 78. doi:10.3847/0004-637X/822/2/78

Jullo, E., Kneib, J.-P., Limousin, M., et al. 2007, New Journal of Physics, 9, 447. doi:10.1088/1367-2630/9/12/447

Jullo, E. & Kneib, J.-P. 2009, MNRAS, 395, 1319. doi:10.1111/j.1365-2966.2009.14654.x

Karman, W., Caputi, K. I., Caminha, G. B., et al. 2017, A&A, 599, A28. doi:10.1051/0004-6361/201629055

Kneib, J.-P., Ellis, R. S., Smail, I., et al. 1996, ApJ, 471, 643. doi:10.1086/177995

Lagattuta, D. J., Richard, J., Clément, B., et al. 2017, MNRAS, 469, 3946. doi:10.1093/mnras/stx1079

Le Fèvre, O., Saisse, M., Mancini, D., et al. 2003, Proc. SPIE, 4841, 1670. doi:10.1117/12.460959

Limousin, M., Kneib, J.-P., & Natarajan, P. 2005, MNRAS, 356, 309. doi:10.1111/j.1365-2966.2004.08449.x

Lotz, J. M., Koekemoer, A., Coe, D., et al. 2017, ApJ, 837, 97. doi:10.3847/1538-4357/837/1/97

Mahler, G., Richard, J., Clément, B., et al. 2018, MNRAS, 473, 663. doi:10.1093/mnras/stx1971

Meštrić, U., Vanzella, E., Zanella, A., et al. 2022, MNRAS, 516, 3532. doi:10.1093/mnras/stac2309

Postman, M., Coe, D., Benítez, N., et al. 2012, ApJS, 199, 25. doi:10.1088/0067-0049/199/2/25

Richard, J., Claeyssens, A., Lagattuta, D., et al. 2021, A&A, 646, A83. doi:10.1051/0004-6361/202039462

Roberts-Borsani, G., Treu, T., Chen, W., et al. 2023, Nature, 618, 480. doi:10.1038/s41586-023-05994-w

Steinhardt, C. L., Jauzac, M., Acebron, A., et al. 2020, ApJS, 247, 64. doi:10.3847/1538-4365/ab75ed

Treu, T., Roberts-Borsani, G., Bradac, M., et al. 2022, ApJ, 935, 110. doi:10.3847/1538-4357/ac8158

Vanzella, E., Castellano, M., Bergamini, P., et al. 2022, ApJ, 940, L53. doi:10.3847/2041-8213/ac8c2d

Vanzella, E., Castellano, M., Meneghetti, M., et al. 2017, ApJ, 842, 47. doi:10.3847/1538-4357/aa74ae

Vanzella, E., Meneghetti, M., Caminha, G. B., et al. 2020, MNRAS, 494, L81. doi:10.1093/mnrasl/slaa041

Vanzella, E., Caminha, G. B., Rosati, P., et al. 2021, A&A, 646, A57. doi:10.1051/0004-6361/202039466

Vanzella, E., Loiacono, F., Bergamini, P., et al. 2023, arXiv:2305.14413. doi:10.48550/arXiv.2305.14413

Strong Gravitational Lensing in the Era of Big Data
Proceedings IAU Symposium No. 381, 2024
H. Stacey, A. Sonnenfeld & C. Grillo, eds.
doi:10.1017/S1743921323004052

Combined strong and weak gravitational lensing mass measurements in clusters of galaxies

Davide Abriola

University of Milan. email: davide.abriola@unimi.it

Abstract. Gravitational lensing in clusters of galaxies is one of the most powerful methods to probe the dark matter mass distribution inside such systems, after mapping the baryonic component, and test the currently accepted ΛCDM cosmological paradigm. With the advent of new high-resolution facilities such as the JWST, strong lensing (SL) is capable of providing extremely accurate mass measurements in the densest regions of such structures. Weak lensing (WL) provides complementary information by measuring the total mass distribution in the outskirts of galaxy clusters, where no multiple images of background sources are produced.

In my talk, I will present updates on the WL mass reconstruction of the Frontier Fields galaxy cluster Abell 2744 ($z = 0.308$) based on Subaru, Magellan, and JWST data, and I will show how the results obtained, combined with accurate SL modeling of this lens, provide a consistent picture of the cluster total mass distribution. I will discuss the pipeline used for this work, the extensive checks performed on the different datasets and the scientific results obtained. Being composed of several substructures undergoing a merging process, the complex geometry of this cluster makes it an ideal laboratory to verify the consistency and reliability of the results obtained with the two methods. I will compare the predictions of the SL models extrapolated in the outer regions of the cluster with the non-parametric WL mass reconstruction to look for potential systematic effects affecting the SL analysis.

Keywords. Cosmology: observations – dark matter – Galaxies: clusters: individual (Abell 2744) – Gravitational lensing: weak

1. Introduction

Galaxy clusters have been proved to be powerful cosmological probes: indeed, being among the most massive systems of the Universe, they constitute the most recent phase of the hierarchical structure formation, and thus represent the ideal laboratory to study the evolution and interaction between galaxies. For several applications, it is crucial to rigorously map their total mass distribution: e.g., once the baryonic component has been mapped, one can infer the physical properties of dark matter and hence shed new light on its nature. These findings can then be compared with the results of cosmological simulations, in order to test the currently accepted ΛCDM structure formation paradigm. Gravitational lensing is, for this purpose, one of the most robust tools we have at our disposal, since, differently from other methods, it does not rely on stringent hypotheses about the physical nature, state or composition of the deflecting mass. Thanks to the extremely high density of their inner cores (up to hundreds of kpc from their centres), galaxy clusters can act as strong gravitational lenses, capable of producing elongated arcs and hundreds of multiple images of background sources that otherwise could not be observed.

If one wants to map the total mass distribution up to the less dense outskirts of a cluster (i.e., up to few Mpc from their cores), where no multiple images are formed, then weak lensing (WL) provides a complementary and independent probe. The statistical study of the slight distortion induced on the shape of background galaxies allows indeed for a robust and model-free reconstruction of the total mass distribution. Hence, combined strong and weak lensing studies allow one to map the projected total mass of the deflectors on different scales, and an excellent agreement between the two probes has been reported in several clusters.

Among the most massive clusters, Abell 2744 (lying at redshift $z = 0.308$) shows one of the most complex merger phenomena ever observed, detected both in the radio and in the X-ray. A recent WL analysis (Medezinski et al. 2016) based on Subaru/Suprime-Cam imaging yielded new evidence to support this picture, by discovering the presence of four substructures undergoing multiple mergers. More recently, thanks to its peculiarity, this system has been the target of still ongoing surveys, including the Beyond Ultra-deep Frontier Fields And Legacy Observations (BUFFALO) survey, as well as the James Webb Space Telescope (JWST) Early Release Science (ERS) program Grism Lens Amplified Survey from Space (GLASS).

In this talk, I will present a new WL analysis of the galaxy cluster Abell 2744 using deep Magellan/MegaCam multi-band imaging data covering a field of view of approximately $31' \times 33'$. The analysis has been carried out with brand-new shape measurement and PSF reconstruction softwares, which we first tested on less deep Subaru/Suprime-Cam data of the same system. We will show the various tests performed to look for possible sources of systematics, in order to obtain a robust WL mass distribution which does not suffer from the dilution due to foreground sources. Finally, we will compare our result with the afore-mentioned WL analysis and a complementary strong lensing (SL) study.

2. A brief recall of WL methodology

We carried out our WL reconstruction starting from the measurements of the shapes of background galaxies, i.e., lying at a higher redshift than the lens (for a review see, e.g., Bartelmann & Schneider 2001). Given these measurements, we first recover the two-dimensional *shear* field γ induced by the gravitational potential of the deflector through a moving average over a pixelized grid. We then reconstruct the total surface mass density distribution Σ of the cluster from γ by applying the procedure first introduced by Lombardi & Bertin (1998): indeed, it can be shown that the gradient of Σ can be expressed as a suitable combination of the derivatives of γ. Therefore we can minimize the difference between the measured γ and such gradient to recover the expected surface mass density.

3. Data description

We performed the WL analysis of the galaxy cluster Abell 2744 by applying a brand-new pipeline for the measurement of the spatial variation of the point spread function (PSF) and the shapes of background galaxies. For this reason, we first tested the analysis softwares on Subaru/Suprime-Cam BR_cz' imaging data of the same cluster, covering a field of view of approximately $24' \times 27'$. After this preliminary study, we analyzed new data collected in 2018 with the MegaCam camera, located at the $f/5$ focus of the 6.5-m Magellan 2 Clay telescope at Las Campanas Observatory, Chile. The observations were carried out in the three filters g, r, and i, and from them an effective $\simeq 31' \times 33'$ composite image was processed. The observation details of both the instruments are listed in Table 1.

Table 1. The specifics of the observations.

Instrument	Filter	Exp. time [10^4 s]	$m_{\rm lim}$ (AB mag)	FWHM [$''$]
	g	2.04	28.7	0.541
MegaCam	r	1.20	28.4	0.487
	i	4.80	27.8	0.520
	B	0.21	27.9	1.16
Suprime-Cam	R_c	0.31	27.7	1.23
	z'	0.36	27.5	0.91

Notes: $m_{\rm lim}$ denotes the 5σ limiting magnitude within a Kron-like aperture. The point spread function (PSF) standard deviation was estimated as the median of the detected star FWHM size distribution. The latter measurements were realized in order to produce a color-composite image of the cluster, on which we performed the analysis.

We first identified the astronomical sources located on the field of view with the software SExtractor (Bertin & Arnouts 1996), and we later classified them either as galaxies or stars by studying their distribution in size and magnitude. Following this method, we identified 695 objects as stars and 86945 as galaxies.

Afterwards, we further distinguished the galaxies between potential cluster members and background ones. To do so, we relied on the procedure outlined by Medezinski et al. (2010, 2018). We studied the distribution of the galaxies in a color-color (CC) space and, for each CC cell, evaluated the mean distance from the cluster centre†. We thus restricted to the region occupied by those galaxies having the smallest distance from the centre (less than 6 Mpc), and assumed them to likely be cluster members. Since the contamination between cluster members and background galaxies represents one of the primary sources of systematic uncertainties in WL, we verified *a posteriori* the purity and completeness of our cluster galaxies sample by comparing it to the deep VLT/MUSE spectroscopic redshift catalog presented by Bergamini et al. (2023a), containing the redshifts of thousands of galaxies located within few arcmins from the BCG. Through this comparison, we achieved an accuracy of $\simeq 93\%$ and a precision of $\simeq 83\%$. Finally, we isolated the foreground galaxies from the background ones by relying once again on the CC-space methodology described above: we thus obtained a catalog consisting of 54614 potentially background galaxies, corresponding to an average number density of $\simeq 54$ galaxies/arcmin^{-2}.

4. Analysis

4.1. *Shape measurement*

The measurement of the shapes of background galaxies represents one of the most crucial steps in WL analysis, since they have to be measured with high accuracy. It is therefore necessary to carefully take into account the effects induced by the PSF, since it introduces an artificial shape distortion, and properly correct for them. This can be done by measuring the shapes of the stars located on the field of view: stars are indeed ideally point-like sources. Therefore, if they appear extended and/or distorted, this is likely due to the PSF: hence, measuring the shapes of observed stars allows one to quantify its effects. We reconstructed the spatial variation of the PSF all over the field of view with the software mccd (Liaudat et al. 2021).

We then proceeded with the measurements of the shapes of the background galaxies with the module ngmix (Sheldon 2015; Sheldon & Huff 2017). At the end of this step,

† assumed to be the south-east brightest cluster galaxy (BCG).

we had to clean our sample by removing galaxies with an high error on the shape measurement, galaxies with a overlapping isophotes, etc. This left us with 13942 background galaxies to compute the mass distribution with, i.e., $\simeq 15$ galaxies/arcmin^2.

We thus reconstructed the shear field γ. To do so, we pixelized the field of view onto a 480 px $\times$ 500 px grid and introduced an isotropic two-dimensional Gaussian weight function with standard deviation equal to $0.5'$. Since background galaxies are distributed in redshift, it was necessary to weigh them according to their distance in a proper manner. To do so, we estimated their redshift distribution by comparing our sample with the photometric redshift catalog centered in the Hubble Deep Field - North (HDF-N) by Yang et al. (2014), after dividing the galaxies per magnitude bin.

4.2. *Mass reconstruction*

Finally, we reconstructed the total surface mass distribution Σ by implementing the method outlined by Lombardi & Bertin (1998). We estimated a statistical error σ_k on this reconstruction through a *reshuffling* procedure (Lombardi & Bertin 1999), i.e., we produced further 120 mass density maps, each time shuffling the coordinates of the background galaxies, and evaluated the standard deviation on the resulting realizations of Σ.

Eventually, we looked for potential systematic effects. Firstly, we evaluated the so-called B-modes† and verified they were consistent with a null signal. Secondly, we took care of the *mass-sheet degeneracy* by estimating the value of the additional constant to add Σ to. We modelled the main density peak identified in our surface mass distribution (the one in the south-east part of the cluster, see later) in terms of a softened pseudo-isothermal elliptical mass distribution (SPIEMD), depending, among the other parameters, on the additive constant. We determined the parameters of the model by means of a Bayesian inference.

5. Results

At a distance of 2 Mpc from the south-east BCG, we recovered a total mass of $(2.5 \pm 0.2) \times 10^{15}\,\mathrm{M}_\odot$, thus making Abell 2744 one of the most massive clusters. To further corroborate the validity of our results, we compared our findings with the SL study by Bergamini et al. (2023b) based on new deep, high-resolution JWST imaging, after extrapolating their best-fit model up to 700 kpc. As can be seen in the *right* panel of Figure 1, where both the best-fit SL model and the WL surface mass distributions are overlaid on a Magellan g-band extract of the field of view, a good agreement emerges. Only a slight discrepancy is present, likely due to the presence of a saturated star in the Magellan images at our disposal. As a further test, we compared the cumulative radial mass profiles emerging from both the probes, that are displayed in Figure 2. As can be seen, our findings are in a nice agreement within one σ_k.

Our analysis reveals the presence of three density peaks (having a S/N greater than 7), all of which lying in correspondence of those observed in the SL surface mass distribution from Bergamini et al. (2023b). A slight overlap with the locations of the substructures detected in the aforementioned WL study by Medezinski et al. (2016) (the *blue* crosses in the *left* panel of Figure 1) is also observed. In particular, the south-east sub-structure detected by the authors of the study lies well within our density peak, as well as the

† Indeed, the two components of the complex shear can be decomposed into two parts, the so-called $E-$ (curl$-$free) and the $B-$ (divergence$-$free) modes. In the context of WL, the latter is identically zero, therefore the presence of non-null $B-$signal can thus be used as a test for the presence of systematic effects (e.g., for a review see Umetsu 2020). To do so, we evaluated Σ after rotating each component of the shear of an angle $\pi/4$ and estimated the rms error on the distribution obtained this way by means of a similar reshuffling procedure.

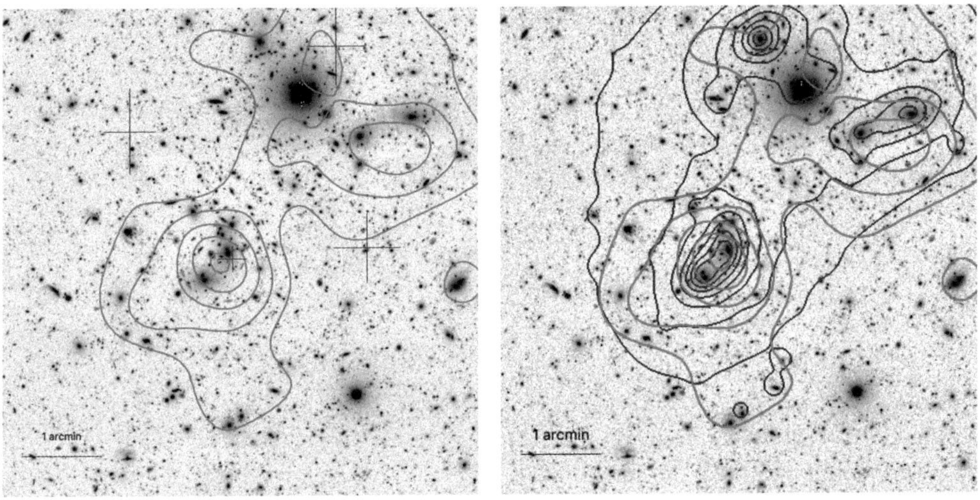

Figure 1. The central $6' \times 6'$ region of the Magellan/MegaCam g-band image with overlaid the contour levels of our surface mass distribution, in *magenta*. *Left*: the *blue* crosses denote the density peaks identified by Medezinski et al. (2016). *Right*: overlaid are also the contour levels of the surface mass distribution emerging from the SL study (Bergamini et al. 2023b, *blue*), referred to the same reference values.

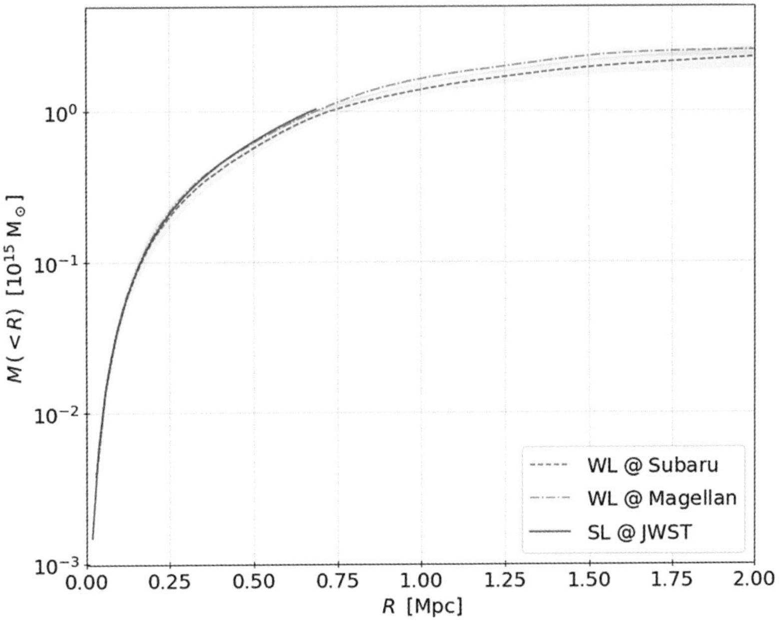

Figure 2. The cumulative radial total mass profile of the cluster in log-scale as obtained with Subaru/Suprime-Cam (blue), on which we first tested the pipeline described in the text, and Magellan/Megacam (red) imaging. Also shown is the cumulative profile (green) in the core of the cluster as a result of the SL analysis based on JWST (Bergamini et al. 2023b). The shaded area denotes the error band corresponding to one rms.

north-west one. This result suggests that the cluster is not relaxed, but is undergoing a complex merging process, as supported by previous papers.

References

Bartelmann, M. & Schneider, P., 2001, Physics Reports, 340, 291
Bergamini, P., Acebron, A., Grillo, C. et al., 2023, A&A, 670, A60
Bergamini, P., Acebron, A., Grillo, C. et al., 2023, arXiv:2303.10210
Bertin, E. & Arnouts, S., 1996, A& ASuppl. Ser., 117, 393
Liaudat, T., Bonnin, J., Starck, J. L., et al., 2021, A&A, 646, A27
Lombardi, M. & Bertin, G., 1998, A&A, 330, 791
Lombardi, M. & Bertin, G., 1999, A&A, 348, 38
Medezinski, E., Broadhurst, T., Umetsu, K., et al., 2010, MNRAS, 405, 257
Medezinski, E., Umetsu, K., Okabe, N., et al., 2016, ApJ, 817, 24
Medezinski, E., Oguri, M., Nishizawa, A. J., et al., 2018, PASJ, 70, 30
Sheldon, E. & Huff, E. M., 2017, ApJ, 841, 24
Sheldon, E., 2015, Astrophysics Source Code Library, record ascl:1508.008
Umetsu, K., 2020, A&A Rev., 28, 7
Yang, G., Xue, Y. Q., Luo, B., et al., 2014, ApJS, 215, 27

Strong Gravitational Lensing in the Era of Big Data
Proceedings IAU Symposium No. 381, 2024
H. Stacey, A. Sonnenfeld & C. Grillo, eds.
doi:10.1017/S1743921323004064

Simulating high-realistic galaxy scale strong lensing in galaxy clusters to train deep learning methods

G. Angora[1,2], P. Rosati[1,3,9], M. Meneghetti[3], M. Brescia[2,4,12], A. Mercurio[2,10], C. Grillo[5,6], P. Bergamini[3,5], A. Acebron[5,6], G. Caminha[7,8], L. Tortorelli[11], L. Bazzanini[1,3] and E. Vanzella[3]

[1]Dipartimento di Fisica e Scienze della Terra, Università di Ferrara, Via Saragat 1, I-44122 Ferrara, Italy. email: gius.angora@gmail.com

[2]INAF – Osservatorio Astronomico di Capodimonte, Salita Moiariello 16, I-80131 Napoli, Italy

[3]INAF – OAS, Osservatorio di Astrofisica e Scienza dello Spazio di Bologna, via Gobetti 93/3, I-40129 Bologna, Italy

[4]Dipartimento di Fisica "E. Pancini", Università di Napoli "Federico II", Via Cinthia 21, I-80126 Napoli, Italy

[5]Dipartimento di Fisica, Università di Milano, via Celoria 16, I-20133 Milano, Italy

[6]INAF – IASF Milano, via A. Corti 12, I-20133 Milano, Italy

[7]Technische Universität München, Physik-Department, James-Franck Str. 1, D-85741 Garching, Germany

[8]Max-Planck-Institut für Astrophysik, Karl-Schwarzschild-Str. 1, D-85748 Garching, Germany

[9]INFN, Sezione di Ferrara, Via Saragat 1, I-44122 Ferrara, Italy

[10]Dipartimento di Fisica, Università di Salerno, Via Giovanni Paolo II, 132, I-84084, Fisciano (SA), Italy

[11]University Observatory, Faculty of Physics, Ludwig-Maximilians-Universität München, Scheinerstr. 1, D-81679 Munich, Germany

[12]INFN, Sezione di Napoli, Via Cinthia 21, I-80126 Napoli, Italy

Abstract. Galaxy-galaxy strong lensing in galaxy clusters is a unique tool for studying the subhalo mass distribution, as well as for testing predictions from cosmological simulations. We describe a novel method that simulates realistic lensed features embedded inside the complexity of observed data by exploiting high-precision cluster lens models. Such methodology is used to build a large dataset with which Convolutional Neural Networks have been trained to identify strong lensing events in galaxy clusters. In particular, we inject lensed sources around cluster members using the images acquired by the Hubble Space Telescope. The resulting simulated mock data preserve the complexity of observation by taking into account all the physical components that could affect the morphology and the luminosity of the lensing events. The trained networks achieve a purity-completeness level of $\sim 91\%$ in detecting such events. The methodology presented can be extended to other data-intensive surveys carried out with the next-generation facilities.

Keywords. Gravitational lensing, galaxies: clusters, image processing, deep learning

---

1. Introduction

Strong gravitational lensing proves highly effective in studying the galaxies and galaxy clusters' mass distribution by also acting as a powerful tool to test cosmological models. For example, strong lensing has been used to: (*i*) estimate the Hubble constant exploiting time-delay measurements (e.g. Suyu *et al.* 2017, 2020; Grillo *et al.* 2018; Millon *et al.* 2020; Moresco *et al.* 2022); (*ii*) constrain the dark energy equation of state (e.g. Jullo *et al.* 2010; Cao, Covone, & Zhu 2012; Collett & Auger 2014; Caminha *et al.* 2022); (*iii*) estimate the dark matter fraction in massive early-type galaxies (e.g. Grillo 2010; Tortora *et al.* 2010; Sonnenfeld *et al.* 2015); (*iv*) test cosmological simulations and/or dark matter properties (e.g. Meneghetti *et al.* 2020, 2022); (*v*) analyze galaxy structures and their evolution (e.g. Treu & Koopmans 2002; Auger *et al.* 2010; Sonnenfeld *et al.* 2013); (*vi*) study lensed (magnified) high-redshift sources revealed by the cluster that acts as cosmic telescope (Swinbank *et al.* 2009; Richard *et al.* 2011; Vanzella *et al.* 2020, 2021). In addition, the total mass distribution of clusters can be studied with strong-lensing cluster-scaled models which exploit an increasing number of multiple images of background sources (e.g. Caminha *et al.* 2017, 2019; Acebron *et al.* 2018; Bergamini *et al.* 2019, 2021, 2023; Lagattuta *et al.* 2019, 2022).

In this work, we explore the capabilities of deep learning methods to automatically search for galaxy–galaxy strong lenses (GGSLs) in galaxy clusters. In these environments, due to the large-scale contribution of the cluster, the probability of GGSLs is generally higher than in the field. Although these systems have been identified through visual inspection of candidates (e.g. Le Fèvre & Hammer 1988; Jackson 2008; Sygnet *et al.* 2010; Pawase *et al.* 2014), in recent years deep learning techniques have been proven as an efficient method to solve this problem automatically (e.g. Petrillo *et al.* 2017, 2019; Metcalf *et al.* 2019; Li *et al.* 2020, 2021; Gentile *et al.* 2022; Cañameras *et al.* 2021; Akhazhanov *et al.* 2022), preparing for the upcoming "Big Data Era" in astronomy inaugurated by the launch of the ESA satellite Euclid on July 2023 (Laureijs *et al.* 2014), together with the upcoming surveys carried out with next-generation facilities, such as the Vera Rubin Observatory (Ivezić *et al.* 2019). Indeed, deep learning methods, with their capabilities to search for hidden correlations within high-dimensional datasets, are able to successfully explore this huge amount of data in a few tens of seconds. Beyond this performative result, this work is focused on the main requirement (maybe the most important) for the training of deep learning techniques: the building of a suitable knowledge space (KB) based on which a convolutional neural network (CNN, LeCun 1989, 1998), can detect GGSL in the dense environment of galaxy clusters. Indeed, GGSLs are rare phenomena. For this reason, several efforts have been carried out in the last decades to simulate these events. For example, lens galaxies can be extracted from semi-analytical catalogs (e.g. with the Millennium Observatory project, as done by Metcalf *et al.* 2019, or by Leuzzi *et al.* 2023) or by modeling the mass density profile of photometrically selected galaxies (e.g. Petrillo *et al.* 2017, 2019; Li *et al.* 2020, 2021; Gentile *et al.* 2022; Cañameras *et al.* 2021; Akhazhanov *et al.* 2022). On the other hand, lensed sources can be simulated by describing their surface brightness distributions (e.g. Petrillo *et al.* 2017, 2019; Li *et al.* 2020, 2021; Gentile *et al.* 2022) or directly sampled from observations (e.g. Meneghetti *et al.* 2008, 2010; Metcalf *et al.* 2019) and then co-added to real or synthetic images through ray-tracing techniques (e.g. `GLAMER` Metcalf & Petkova 2014; Petkova *et al.* 2014, `GRAVLENS` Keeton 2001). However, in this work, we exploit a novel approach that simulates GGSLs in galaxy clusters by taking advantage of accurate cluster deflection fields provided by high-precision cluster lens models constructed by Bergamini *et al.* (2019, 2021) and Caminha *et al.* (2019). These models consider the cluster-scale and the subhalo (i.e. the cluster members, CLMs) mass components for a given distribution of background sources. Thus, the GGSL simulation is driven by a realistic description of the CLMs, which behaves as though they actually were the lenses in lensing events, since the process takes into account all the physical components that could affect the morphology, the brightness, and the frequency of the events.

As aforementioned, this work focuses on the simulation methodology. All the details about the network implementations, the handling of train-test set splitting, together with a full description of the CNN performances and the analysis of the results can be found in Angora *et al.* (2020) and Angora *et al.* (2023). Thus, the paper extensively emphasizes the simulation process and the KB building in Section 2, by just summarizing the CNN results in Section 3; finally, we draw out conclusions in Section 4 and outline the future perspectives in Section 5.

Throughout the paper, we adopt a flat ΛCDM cosmological model with Ω_M=0.3, Ω_Λ= 0.7, and $H_0 = 70\,\mathrm{km\,s^{-1}\,Mpc^{-1}}$. All astronomical images are oriented north to the top and east to the left. Unless otherwise specified, magnitudes are in the AB system.

2. Simulating GGSLs in Hubble images: building the KB

The simulation exploits the deflection angle maps of eight galaxy clusters obtained from cluster lens models† provided by Bergamini *et al.* (2019, 2021) and Caminha *et al.* (2019). The cluster sample spans a redshift range between 0.234 and 0.587 (it is fully described in Table 1 of Angora *et al.* 2023). A complete description of the cluster-scale lensing modeling can be found in Bergamini *et al.* (2021), here we summarize the main points. The gravitational potential of each cluster is described through a parametric model which includes: (*i*) the cluster-scale components, i.e. dark matter halo and the smooth intracluster hot-gas mass (when available, e.g. Bonamigo *et al.* 2017, 2018), (*ii*) the clumpy terms associated to the CLMs; (*iii*) any further massive objects in the cluster outskirts or along the line of sight (LoS). We underline that the mass density profile of the subhalos is modeled with a circular, singular dual-pseudo isothermal profile (Limousin, *et al.* 2005; Elíasdóttir *et al.* 2007), calibrated by measuring stellar velocity dispersions of a large sample of CLMs. Such lens models reproduce the observed positions of many multiple images (ranging from ~ 20 to ~ 200) with a typical accuracy of $\lesssim 0.5''$.

In this work, we inject GGSLs in three images acquired by the Hubble Space Telescope (HST) Advanced Camera for Survey (ACS), as part of the CLASH (Cluster Lensing And Supernova survey with Hubble, Postman *et al.* 2012) and HFF (Hubble Frontier Fields, Lotz *et al.* 2017) surveys. In particular, we exploit three ACS bands: F435W, F606W, and F814W. However, this process can be applied to any image, acquired with any kind of instrument as long as the related deflection field is available.

The simulation process is carried out by using a python package, `PyLensLib` (Meneghetti 2022). It is graphically depicted in Figure 1, and summarizes as follows.

(0) Inputs: (*i*) the source, described as a Sérsic profile (Sérsic 1963, 1968) and with a star-forming galaxy spectral energy distribution (SED, Kinney *et al.* 1996), together with its intrinsic (F814W) magnitude and its redshift ‡; (*ii*) the angle deflection maps (computed from the lens model given the source and lens redshift), $\vec{\alpha}$, which describes the relation between the source real position ($\vec{\beta}$) and the observed position $\vec{\theta}$ (see also right panels in Figure 1), through the lens equation:

$$\vec{\beta} = \vec{\theta} - \vec{\alpha} \tag{1}$$

† The cluster lens models are publicly available at `https://www.fe.infn.it/astro/lensing/`.
‡ The adopted ranges of the Sérsic parameters can be found in Table 2 of Angora *et al.* (2023).

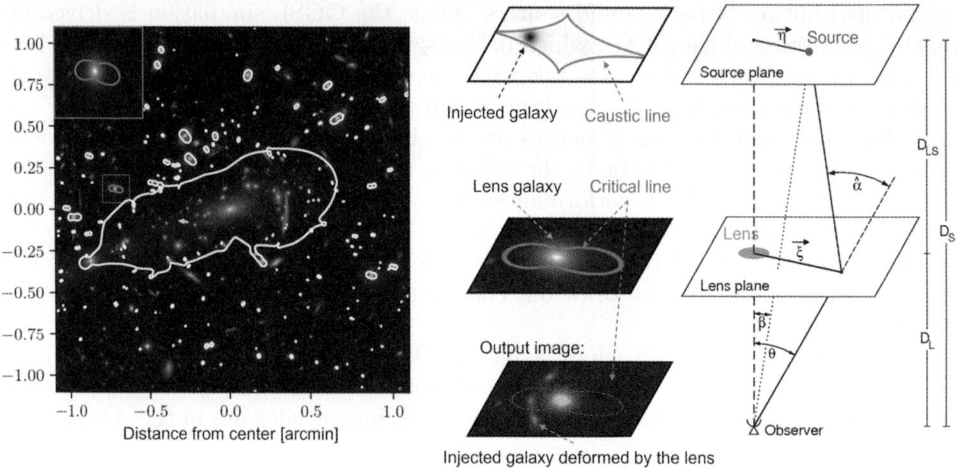

Figure 1. A flow chart representing the simulation process. *Left panel*: the cluster MACSJ 1206 with the critical lines (in white) at $z = 2.5$ from the lens model built by Bergamini *et al.* (2019), where a specific secondary critical line is selected (marked in red and zoomed-in the $10''$ inset axis). *Central panels*: the source is injected inside the caustic line corresponding to the selected critical line; the simulated GGSL is obtained through ray-tracing. *Right panels*: the corresponding geometric description of the lensing event.

(1) The critical lines (on the lens plane, see the left panel in Figure 1) are computed from the deflection field by imposing that the magnification goes to infinity (the elements of the magnification tensor can be computed from the convergence and shear maps, i.e. the Jacobian matrix describing the image deformation).
(2) a secondary critical line is randomly extracted inside an (equivalent) Einstein radius range: $\theta_E \in (0.2'', 3.5'')$ (prevent the sampling of the primary critical lines and very small-scaled galaxies). These limits are compatible with the expected Einstein radii distribution (see Figure 7 in Meneghetti *et al.* 2022).
(3) The selected secondary critical line is mapped into the corresponding caustic line (on the source plane) using the lens equation 1 (see central panels in Figure 1). The source is injected within a buffer surrounding the caustic (whose width is a fraction of the source effective radius, $0.5r_e$).
(4) Since the lens mapping preserves the surface brightness, i.e. $I(\vec{\theta}) = I(\vec{\beta})$, the resulting surface brightness on the lens plane is computed as $I(\vec{\theta} = \vec{\beta} + \vec{\alpha})$. Finally, the generated lensed system is convolved with the HST point spread function (PSF) and coadded to the HST image. In this work, the PSF's are estimated with `morphofit` (Tortorelli & Mercurio 2023).

In order to reproduce an observed Universe, the source intrinsic magnitude and redshift are extracted by exploiting the COSMOS 2015 catalog (Laigle *et al.* 2016) complemented with HST Deep Field North and South observations (Williams *et al.* 1996; Metcalfe *et al.* 2001). Specifically, we estimate the number counts (i.e. the number of galaxies per squared degree per magnitude bin) in F814W down to 29 mag, and we fit a redshift probability density function (PDF) in 6 magnitude bin, i.e. $p(z|\Delta m) = az^2 e^{-(z/z_0)^b}$ (Lombardi & Bertin 1999; Lombardi *et al.* 2005, see also Figure 3 in Angora *et al.* 2023). In this way, we can extract a source magnitude (in the range $22 - -29$) and, given this magnitude, a corresponding redshift. In the COSMOS 2015 catalog the photometric redshift limit is ~ 7, which is suitable for our experiments since the reddest band is the

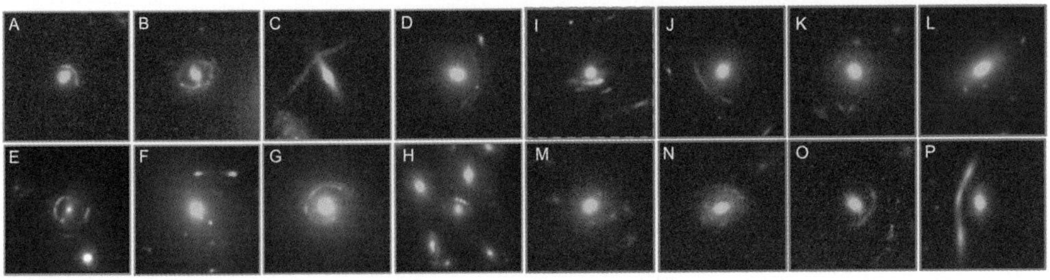

Figure 2. Comparison between real (in the left panels, with green edges) and simulated (in the right panels, with red edges) GGSLs. Note that in the panel *I* there is a combination of real and simulated events: the blue arc is a true GGSL (with a source at $z \sim 2$), whereas the red arc is injected ($z \sim 6$). *References.* Desprez *et al.* (2018): panels A, B, D, E and G; Diego *et al.* (2015): panel C; Smith *et al.* (2005): panels F and H, Caminha et al. (2016): panel I.

F814W. On the other hand, we set a minimum value for the source redshift equals to $z_{src} = z_{cls} + 0.4$, which is suggested by studying the lensing cross-section of galaxies in galaxy clusters (Meneghetti *et al.* 2022). Figure 2 shows a comparison between real and simulated GGSLs.

Finally, the building of KB is completed by selecting a pure sample of non-GGSL, i.e. the negative class in the dataset. Lensing experts in our group visually inspected thousand of cutouts centered on CLM centroids by excluding any thumbnail that contained strong lensing features. This result has been carried out by exploiting a large spectroscopic coverage part of the CLASH-VLT VIMOS (Rosati *et al.* 2014) program combined with MUSE archival observations (Bacon *et al.* 2012, 2014, 2015). The resulting non-GGSL sample comprises of ~ 1500 objects. Therefore, the GGSL set in the KB is composed of ~ 3000 events, i.e. two times the number of non-GGSLs, with the aim to ensure a large variety of lensing mock examples. This intentional mismatch is balanced during the augmentation pre-processing phase, which leads to a balanced training set of ~ 3800 images for each class. Cutouts have a side of 128×128 pixels (i.e. $3.84''$ by side).

3. Classification performances

The CNN classification performances have been measured with a set of metrics that can be computed from the binary confusion matrix: the average efficiency (*AE*), purity (*pur*), completeness (*compl*), and the *F1* - score. These estimators can completely describe the CNN capability to disentangle the GGSLs from the negative class. In particular, the AE is a global score measured by combining both classes, while the others refer to each class: *pur* is related to the degree of contamination in the network predicted sample, *compl* is the capability to identify *all* the sources, the *F1* is the harmonic mean between purity and completeness, which has been included to take into account the typical dichotomy between these estimators.

We tested several CNN architectures; here, we summarize the results achieved by the two best networks. The first is inspired by the work of Simonyan & Zisserman (2014), a VGG-like (Visual Group Geometry) network. The second architecture was implemented based on a posterior analysis of the results, it consists of three parallel networks, each of them processes a single channel, for this reason, we named it SC-VGG. All the details about these networks can be found in Angora *et al.* (2020, 2023).

Table 1 compares the classification performances of the two networks in terms of the adopted statistical estimators. Moreover, we tested the CNN capabilities to identify GGSLs by exploiting just a band (last three columns in Table 1). Globally speaking, all the studied configurations are purer than complete with respect to the identification

Table 1. Classification performance comparison.

	[%]	VGG	SC-VGG	*F435W*	*F606W*	*F814W*
	AE	87.7	89.4	87.2	86.1	86.8
	pur	93.4	93.1	91.8	91.1	91.5
GGSL	*compl*	88.6	91.7	89.8	88.8	89.3
	F1	91.0	92.4	90.8	89.9	90.4
	pur	76.7	81.4	77.5	75.4	76.6
non-GGSL	*compl*	85.4	84.1	81.3	79.9	80.9
	F1	81.1	82.8	79.3	77.6	78.7

Notes: Beyond the comparison between the two architectures, we show the results achieved by networks trained with a single band (last three columns).

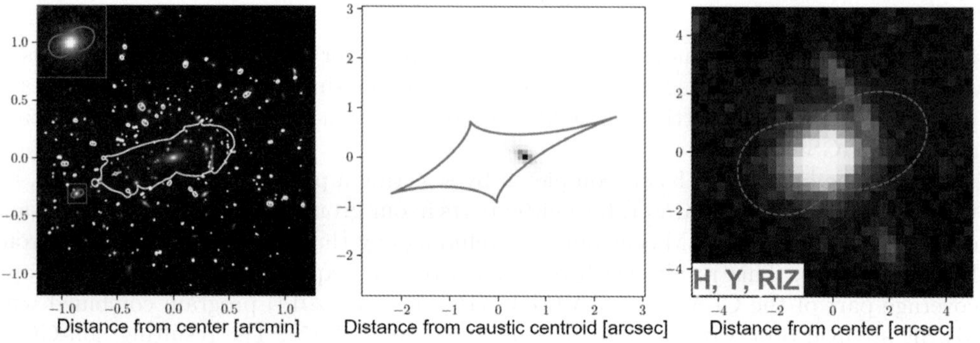

Figure 3. Simulation process in an euclidized image. *Left panel*: the euclidized FoV of the cluster MACS J1206, together with the critical lines. The selected secondary critical line is in red and is also shown in the upper-left inset axis. *Central panel*: the corresponding caustic (on the source plane at $z = 1.5$) with the source ($\mathrm{mag_{RIZ}} = 23.0$). *Right panel*: the resulting GGSLs as a color-composite image.

of GGSLs, with purity measures ranging in $91.1 - 93.4$; whereas the completeness fluctuation range is lower ($88.6 - 91.7$). Opposite behavior regarding the non-GGSL class. The SC-VGG shows more stable results for both classes by achieving the highest *F1*-scores. This is due to its capability to identify better high-z GGSLs since it is not affected by the dropout problem (see the discussion in Angora *et al.* 2023).

Interestingly, we find that the network is able to classify GGSLs by also using just one band, even if with performance lower than those achieved by combining the three filters. It is not surprising that the best performances are carried out by using the bluer band (F435W). Although this filter has a lower signal-to-noise ratio, the network exploits the higher contrast between the lens (redder) and the lensed source (typically bluer).

4. Conclusion

In this work we outline the methodology with which a large dataset can be built to train CNNs in order to identify GGSL in galaxy clusters. The presented simulation process takes advantage of high-precision cluster deflection fields, which have been used to generate high-realistic mock lensed features embedded in the cluster environment. The resulting dataset reproduces the complexity of the lensing phenomenon by considering all the physical components that characterize the lenses. The implemented CNNs are able to classify GGSLs with a purity-completeness trade-off $\sim 91\%$.

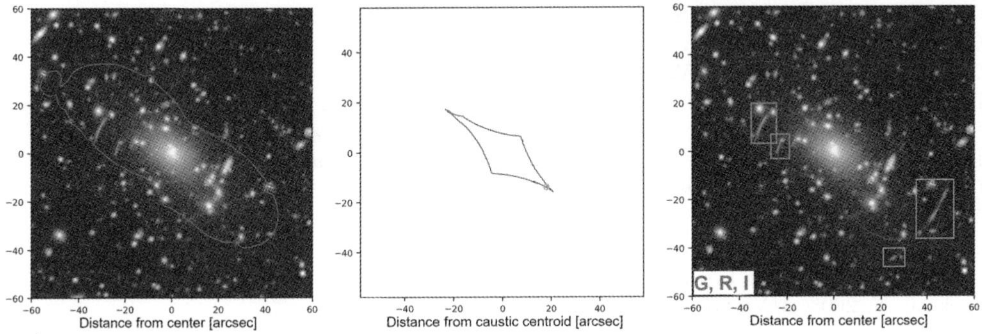

Figure 4. Cluster-scale simulation process in VST simulated image. *Left panel*: the VST FoV of the cluster RX J2248, together with the primary critical lines. *Central panel*: the corresponding caustic (on the source plane at $z = 2.9$) where the source location is shown as a yellow circle ($\mathrm{mag_{RIZ}} = 25.0$). *Right panel*: the resulting simulated arcs (in yellow boxes) as a color-composite image. As comparison, two real arcs are enclosed by green boxes.

5. Future perspectives

This methodology has been tested with images acquired by the HST. Currently, we are extending this process to other instruments. For example, Figure 3 shows a simulated GGSL injected in the *euclidized* version of the HST FoV of the cluster MACS J1206. The euclidization has been carried out with the tool `HST2EUCLID` (Bergamini et al. in prep.).

In addition, the same process can be exploited to simulate extended arc-like structures on a cluster scale. We are simulating thousands of arcs by injecting sources near the primary critical line. We plan to search for extended arcs in images acquired by the VLT Survey Telescope (VST). In order to exploit the high-precision cluster deflection maps (whose cluster lens models have been built with HST images), we implemented a tool that transforms HST images into VST mock data. The simulation process (graphically represented in Figure 4) is similar to that described in this work. In particular, in order to obtain high-magnified lensed arcs, we select the regions where the magnification is higher than 70. The right panel in Figure 4 shows a comparison between the injected arc (inside yellow boxes) and two observed arcs (inside green boxes).

References

Acebron, A., Cibirka, N., Zitrin, A., *et al.* 2018, *Apj*, 858, 42.
Akhazhanov, A., More, A., Amini, A., *et al.* 2022, *MNRAS*, 513, 2407.
Angora, G., Rosati, P., Brescia, M.,*et al.* 2020, *A&A*, 643, A177.
Angora, G., Rosati, P., Meneghetti, M., *et al.* 2023, *A&A*, 676, A40.
Auger, M.W., Treu, T., Gavazzi, R., *et al.* 2010, *Apj*, 721, L163.
Bacon, R., Accardo, M., Adjali, L., *et al.* 2012, *The Messenger*, 147, 4.
Bacon, R., Vernet, J., Borisova, E., *et al.* 2014, *The Messenger*, 157, 13.
Bacon, R., Brinchmann, J., Richard, J., *et al.* 2015, *A&A*, 575, A75.
Bergamini, P., Rosati, P., Mercurio, A., *et al.* 2019, *A&A*, 631, A130.
Bergamini, P., Rosati, P., Vanzella, E., *et al.* 2021, *A&A*, 645, A140.
Bergamini, P., Agnello, A., and Caminha, G.B. 2021, *A&A*, 648, A123.
Bergamini, P., Acebron, A., Grillo, C., *et al.* 2023, *Apj*, 952, 84.
Bonamigo, M., Grillo, C., Ettori, S., *et al.* 2017, *Apj*, 842, 132.
Bonamigo, M., Grillo, C., Ettori, S., *et al.* 2018, *Apj*, 864, 98.
Caminha, G.B., Grillo, C., Rosati, P., *et al.* 2016, *A&A*, 587, A80.
Caminha, G.B., Grillo, C., Rosati, P., *et al.* 2017, *A&A*, 607, A93.
Caminha, G.B., Rosati, P., Grillo, C., *et al.* 2019, *A&A*, 632, A36.
Caminha, G.B., Suyu, S.H., Grillo *et al.* 2022, *A&A*, 657, A83.

Cañameras, R., Schuldt, S., Shu, Y., *et al.* 2021, *A&A*, 653, L6.
Cao S., Covone G., Zhu Z.H. 2012, *ApJ*, 755, 31.
Collett, T.E. and Auger, M.W. 2014, *MNRAS*, 443, 969.
Desprez, G., Richard, J., Jauzac, M., *et al.* 2018, *MNRAS*, 479, 2630.
Diego, J.M., Broadhurst, T., Benitez, N., *et al.* 2015, *MNRAS*, 449, 588.
Elíasdóttir, Á., Limousin, M., Richard, J., *et al.* 2007, *arXiv e-prints*, arXiv:0710.5636.
Gentile, F., Tortora, C., Covone, G., *et al.* 2022, *MNRAS*, 510, 500.
Grillo, C. 2010, *Apj*, 722, 779.
Grillo, C., Rosati, P., Suyu, S.H., *et al.* 2018, *ApJ*, 860, 94.
Ivezić, Ž., Kahn, S.M., Tyson, J.A., *at al.* 2019, *Apj*, 873, 111.
Jackson, N. 2008, *MNRAS*, 389, 1311.
Jullo, E., Natarajan, P., Kneib, J.-P., *et al.* 2010, *Science*, 329, 924.
Keeton, C.R. 2001, *arXiv e-prints*, astro-ph/0102340.
Kinney, A.L., Calzetti, D., Bohlin, R.C. *et al.* 1996, *Apj*, 467, 38.
Lagattuta, D.J., Richard, J., Bauer, F.E., *et al.* 2019, *MNRAS*, 485, 3738.
Lagattuta, D.J., Richard, J., Bauer, F.E., *et al.* 2022, *MNRAS*, 514, 497.
Laigle, C., McCracken, H.J., Ilbert, O., *et al.* 2016, *Apjs*, 224, 24.
Laureijs, R., Hoar, J., Buenadicha, G., *et al.* 2014, *Astronomical Data Analysis Software and Systems XXIII*, 485, 495.
LeCun, Y., Boser, B., Denker, J. S., *et al.* 1989, *Neural Comput*, 1 (4), 541–551.
Lecun, Y., Bottou, L., Bengio, Y., *et al.* 1998, *Proceedings of the IEEE* 86, 11, 2278–2324.
Le Fèvre, O. and Hammer, F.: 1988, *Apj*, 333, L37.
Leuzzi, L., Meneghetti, M., Angora, G., and Euclid Collaboration 2023, *arXiv e-prints*, arXiv:2307.08736.
Li, R., Napolitano, N.R., Tortora, C., Spiniello, C., *et al.* 2020, *Apj*, 899, 30.
Li, R., Napolitano, N.R., Spiniello, C., *et al.* 2021, *Apj*, 923, 16.
Limousin, M., Kneib, J.-P., and Natarajan, P. 2005, *MNRAS*, 356, 309.
Lombardi, M. and Bertin, G. 1999, *A&A*, 342, 337.
Lombardi, M., Rosati, P., Blakeslee, J.P., *et al.* 2005, *Apj*, 623, 42.
Lotz, J.M., Koekemoer, A., Coe, D., *et al.* 2017, *Apj*, 837, 97.
Meneghetti, M., Melchior, P., Grazian, A., *et al.* 2008, *A&A*, 482, 403.
Meneghetti, M., Rasia, E., Merten, J., *et al.* 2010, *A&A*, 514, A93.
Meneghetti, M., Davoli, G., Bergamini, P., *et al.* 2020, *Science*, 369, 1347.
Meneghetti, M. 2022, *Springer*, ISBN: 3-030-73582-6.
Meneghetti, M., Ragagnin, A., Borgani, S., *et al.* 2022, *A&A*, 668, A188.
Metcalfe, N., Shanks, T., Campos, A., *et al.* 2001, *MNRAS*, 323, 795.
Metcalf, R.B. and Petkova, M. 2014, *MNRAS*, 445, 1942.
Metcalf, R.B., Meneghetti, M., Avestruz, C., *et al.* 2019, *A&A*, 625, A119.
Millon, M., Galan, A., Courbin, F., *et al.* 2020, *A&A*, 639, A101.
Moresco, M., Amati, L., Amendola, L., *et al.* 2022, *Living Reviews in Relativity*, 25, 6.
Pawase, R.S., Courbin, F., Faure, C., *et al.* 2014, *MNRAS*, 439, 3392.
Petkova, M., Metcalf, R.B., and Giocoli, C. 2014, *MNRAS*, 445, 1954.
Petrillo, C.E., Tortora, C., Chatterjee, S., *et al.* 2017, *MNRAS*, 472, 1129.
Petrillo, C.E., Tortora, C., Vernardos, G., *et al.* 2019, *MNRAS*, 484, 3879.
Postman, M., Coe, D., Benítez, N., *et al.* 2012, *Apjs*, 199, 25.
Richard, J., Kneib, J.-P., Ebeling, H., *et al.* 2011, *MNRAS*, 414, L31.
Rosati, P., Balestra, I., Grillo, C., *et al.* 2014, *The Messenger*, 158, 48.
Sérsic, J.L. 1963, *Boletin de la Asociacion Argentina de Astronomia La Plata Argentina*, 6, 41.
Sérsic, J.L. 1968, *Cordoba, Argentina: Observatorio Astronomico, 1968.*
Simonyan, K. and Zisserman, A. 2014, *arXiv e-prints*, arXiv:1409.1556.
Smith, G.P., Kneib, J.-P., Smail, I., *et al.* 2005, *MNRAS*, 359, 417.
Sonnenfeld, A., Treu, T., Gavazzi, R., *et al.* 2013, *Apj*, 777, 98.
Sonnenfeld, A., Treu, T., Marshall, P.J., *et al.* 2015, *Apj*, 800, 94.
Suyu, S.H., Bonvin, V., Courbin, F., *et al.* 2017, *MNRAS*, 468, 2590.

Suyu, S.H., Huber, S., Cañameras, R., *et al.* 2020, *A&A*, 644, A162.
Swinbank, A.M., Webb, T.M., Richard, J., *et al.* 2009, *MNRAS*, 400, 1121.
Sygnet, J.F., Tu, H., Fort, B., *et al.* 2010, *A&A*, 517, A25.
Tortora, C., Napolitano, N.R., Romanowsky, A.J., and Jetzer, P. 2010, *Apj*, 721, L1.
Tortorelli, L. and Mercurio, A. 2023, *Frontiers in Astronomy and Space Sciences*, 10, 51.
Treu, T. and Koopmans, L.V.E. 2002, *Apj*, 575, 87.
Vanzella, E., Meneghetti, M., Caminha, G.B., *et al.* 2020, *MNRAS*, 494, L81.
Vanzella, E., Caminha, G.B., Rosati, *et al.* 2021, *A&A*, 646, A57.
Williams, R.E., Blacker, B., Dickinson, M., *et al.* 1996, *Apj*, 112, 1335.

Strong Gravitational Lensing in the Era of Big Data
Proceedings IAU Symposium No. 381, 2024
H. Stacey, A. Sonnenfeld & C. Grillo, eds.
doi:10.1017/S1743921323004039

Shapes and Centroids of 39 Strong Lensing Galaxy Clusters from the Sloan Giant Arcs Survey

Raven Gassis[1], **Matthew B. Bayliss**[1], **Keren Sharon**[2], **Guillaume Mahler**[3], **Michael D. Gladders**[4], **Håkon Dahle**[5], **Michael K. Florian**[6], **Jane R. Rigby**[7], **Lauren Elicker**[1] **and M. Riley Owens** [1]

[1]Department of Physics, University of Cincinnati. email: gassismr@mail.uc.edu

[2]Department of Astronomy, University of Michigan,

[3]Centre for Extragalactic Astronomy, Durham University,

[4]Department of Astronomy and Astrophysics/Kavli Institute for Cosmological Physics, University of Chicago,

[5]Institute of Theoretical Astrophysics, University of Oslo,

[6]Steward Observatory,

[7]University of Arizona, Observational Cosmology Lab, Code 665, NASA Goddard Space Flight Center

Abstract. Strong lensing galaxy clusters provide a powerful observational test of Cold Dark Matter (CDM) structure predictions derived from simulation. Specifically, the shape and relative alignments of the dark matter halo, stars, and hot intracluster gas tells us the extent to which theoretical structure predictions hold for clusters in various dynamical states. We measure the position angles, ellipticities, and locations/centroids of the brightest cluster galaxy (BCG), intracluster light (ICL), the hot intracluster medium (ICM), and the core lensing mass for a sample of strong lensing galaxy clusters from the SDSS Giant Arcs Survey (SGAS). We use iterative elliptical isophote fitting methods and GALFIT modeling on HST WFC3/IR imaging data to extract ICL and BCG information and use `CIAO's Sherpa` modeling on Chandra ACIS-I X-ray data to make measurements of the ICM. Using this multicomponent approach, we attempt to constrain the physical state of these strong lensing clusters and evaluate the different observable components in terms of their ability to trace out the gravitational potential of the cluster.

Keywords. Strong Lensing Galaxy Clusters (SLGCs), Brightest Cluster Galaxy (BCG), Intracluster Medium (ICM), Intracluster Light (ICL)

1. Introduction

An important consequence of Λ Cold Dark Matter (ΛCDM) physics is that galaxy clusters form by the hierarchical formation of DM halo systems. In this scenario, the cluster gradually accretes and incorporates other haloes, eventually forming one large mass halo system that defines the cluster (Beers and Geller 1983).

In a theoretical system only subject to gravitational forces, the brightest cluster galaxy (BCG), intracluster light (ICL), and the hot intracluster medium (ICM) of a given galaxy cluster should align with the dark matter halo of the cluster defined by the core lensing mass (Sastry 1968; Binggeli 1982; Van Den Bosch et al. 2005; Hashimoto et al. 2008; Niederste-Ostholt et al. 2010; Biernacka et al. 2015; Donahue et al. 2016; Wang et al. 2018).

Though there is broad agreement with the predicted alignment, previous work has found that some of the mass components are not always aligned with the dark matter halo or with each other (Van Den Bosch et al. 2005; Sanderson et al. 2009; Skibba et al. 2011; Zitrin et al. 2012; Hikage et al. 2013; Lauer et al. 2014; Martel et al. 2014; Oliva-Altamirano et al. 2014; Wang et al. 2014; Hoshino et al. 2015; Rossetti et al. 2016; Lange et al. 2018; Lopes et al. 2018; Zenteno et al. 2020; De Propris et al. 2021). This suggests that there are mechanisms by which misalignments can occur in the context of ΛCDM or potentially beyond ΛCDM.

Given enough time to relax, the BCG and ICL should revert to their shared orientation and centroid with their dark matter halo (Montes and Trujillo 2018; Wittman et al. 2019). However, work by Kim et al. (2017) and Harvey et al. (2017) found that the BCG can still show misalignments ("wobble") in relaxed clusters which is unaccounted for in the cold dark matter description. Though the ICL is not expected to have any residual "wobbling", DM measurements derived from strong lensing mass models allow us to directly test whether the ICL traces out the shape of the DM halo as predicted.

The ICM gas can exhibit extreme misalignments for disturbed clusters due to merger activity. Even for relaxed clusters, hydrodynamical gas oscillations can cause some deviations in the ICM to persist even when the other components of the cluster system have had time to relax (Markevitch et al. 2001; Churazov et al. 2003; Johnson et al. 2012; Harvey et al. 2017).

2. Data and Measurements

In this work, we apply a multicomponent observational approach incorporating measurements of the BCG, ICL, ICM, and core lensing mass derived from strong lensing. We investigate 39 clusters from the Sloan Giant Arcs Survey (SGAS) with corresponding multiband *HST* data and spectroscopic follow up that enabled well constrained strong gravitational lensing models (Sharon et al. 2020, 2022a,b). A subset of 27 clusters have *Chandra* observations that allow us to make measurements on the BCG, ICL, ICM, and core lensing mass simultaneously.

2.1. *BCG/ICL Measurement*

We modeled the BCG using GALFIT in order to eliminate all interfering light. For the ICL, we derived masks for all objects in the field using `SExtractor`. For both the ICL and BCG, we use iterative elliptical isophote fitting applying the Python `lsq-ellipse` function. To differentiate between the BCG and ICL, we conduct a radial profile analysis of our clusters to find the ICL/BCG transition radius to be about 50 kpc which is consistent with 60 ± 40 kpc as measured by Contini et al. (2022).

2.2. *ICM Measurement*

For the ICM gas, we use the publicly available `CIAO` tools for data reduction to process the Chandra X-ray event data. Each processed X-ray image file was point source subtracted, binned by 2", and set to only include events in the broad energy band (0.5-7 keV with an effective energy of 2.3 keV). We modeled the processed ICM distribution using `CIAO's Sherpa` modeling functions and took the outputted best fit as the true parameters.

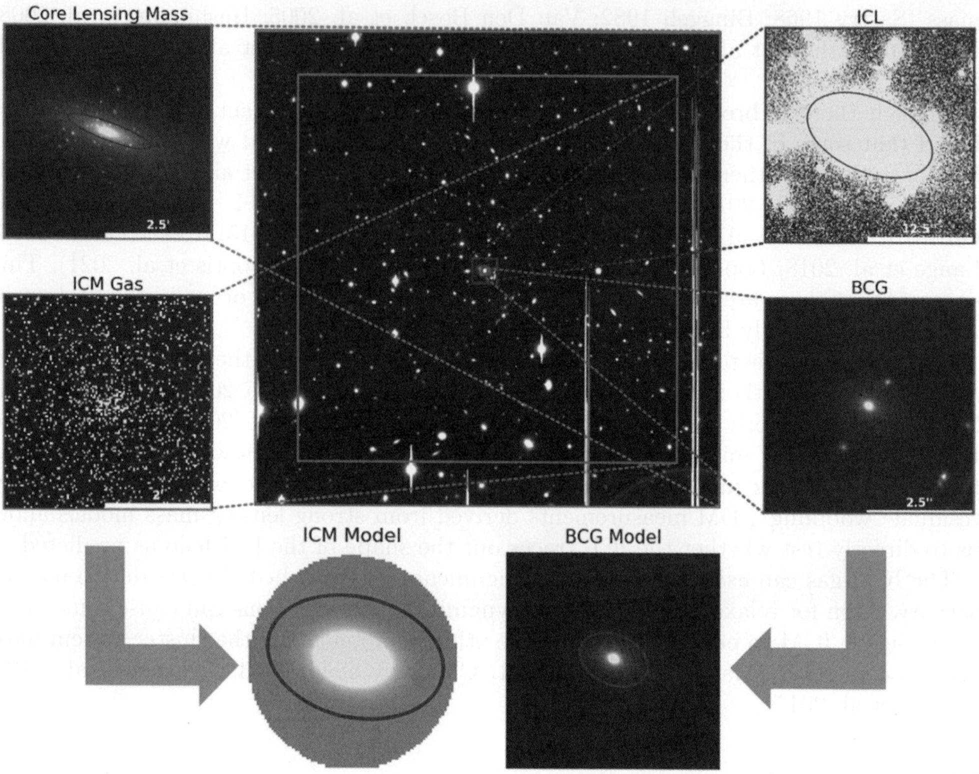

Figure 1. Visualization of the data distributions with the corresponding best fit ellipse for example galaxy cluster J0957p0509. We include a 5"x5" representative color (gri) Subaru image of the field in the center panel for a scale comparison of the different distributions. The ICM and BCG also have visualizations of their models that precede ellipse fitting.

2.3. *Strong Lensing Models and Core Lensing Mass Measurement*

The strong lensing models for these galaxy clusters were derived using the publicly available Lenstool software (Jullo et al. 2007). The lensing models are more fully described in Sharon et al. 2020, 2022a,b. We use 100 lens models drawn from the `Lenstool` MCMC with the derived lensing parameters to sample the posterior probability distribution. In the image plane models, there is an obvious core cluster that dominates the distribution at large scales. This is the core mass distribution we would expect to align with the BCG, ICL, and ICM components. We choose the core lensing mass parameters from the most likely image plane model to be the true values.

3. Results

3.1. *Position Angles*

Figure 2 shows the results of the various cluster components' position angles compared to the position angle of the core lensing mass. Equation 1 defines the difference in position angle.

$$\begin{aligned} \Delta\mathrm{PA} &= |\mathrm{PA}_1 - \mathrm{PA}_2| \qquad (|\mathrm{PA}_1 - \mathrm{PA}_2| \leq 90^\circ) \\ \Delta\mathrm{PA} &= 180 - |\mathrm{PA}_1 - \mathrm{PA}_2| \qquad (|\mathrm{PA}_1 - \mathrm{PA}_2| > 90^\circ) \\ & 0^\circ \leq \Delta\mathrm{PA} \leq 90^\circ \end{aligned} \tag{1}$$

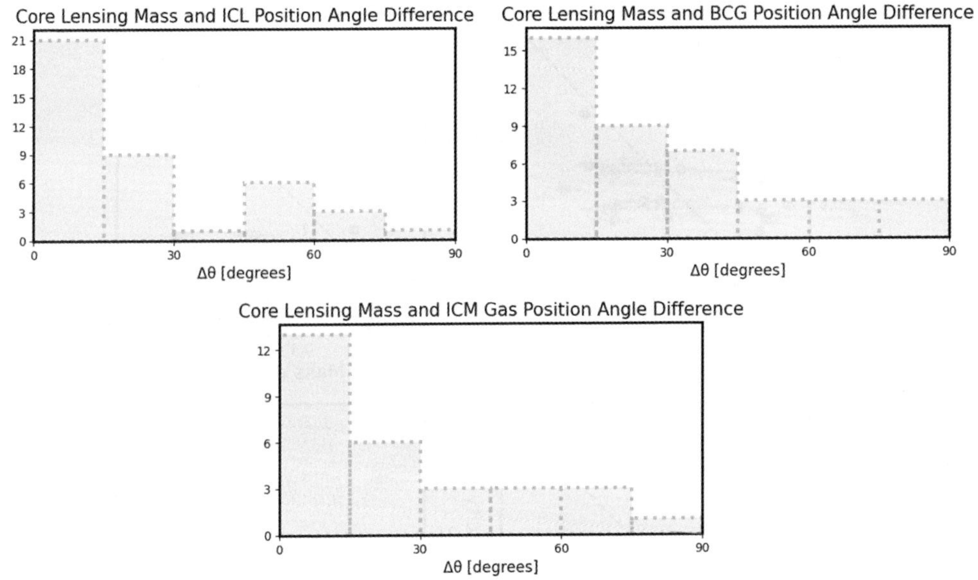

Figure 2. The difference in position angle of the major axis between the core lensing mass and the three other distributions (from top to bottom: ICL, BCG, and ICM Gas)

Generally, we measure small position angle differences which implies that cluster orientation is consistent over a large spatial scale (from a few tens of kpc up to ~1Mpc). This behavior is consistent with previous studies (e.g. Donahue et al. 2016).

However, we still measure multiple large deviations from the expected alignment (ΔPA$> 30°$). The smaller number of deviations between the core lensing mass and ICL suggest that the ICL may be a more viable proxy for the shape and orientation of the dark matter halo distribution since it is gravitationally bound to the core halo as opposed to any specific galaxy member (Niederste-Ostholt et al. 2010; Hashimoto et al. 2014; Donahue et al. 2016).

3.2. *Ellipticities*

Figure 3 compares the ellipticity measurements of the different components of our objects. We use the definition of ellipticity analogous to the flattening parameter defined by $e = 1 - b/a$, where a and b are the semi-major axis and semi-minor axis respectively.

From the top-left panel, we see that, despite some scatter, the dark matter and stellar distribution characterized by the ICL trace out the same shape as predicted by Montes and Trujillo (2018). The bottom 2 panels reveal that the ICM is generally rounder than the dark matter and ICL which reflects the effects of hydrodynamical physics in the ICM (Markevitch and Vikhlinin 2007). In the top-right panel, we see the BCG is circularized due to dynamical friction effects as a consequence of high stellar density (Arena and Bertin 2007).

3.3. *Centroids*

From figure 4, we are able to gather an understanding of the centroid difference of the different components of our galaxy clusters when compared to one another. When

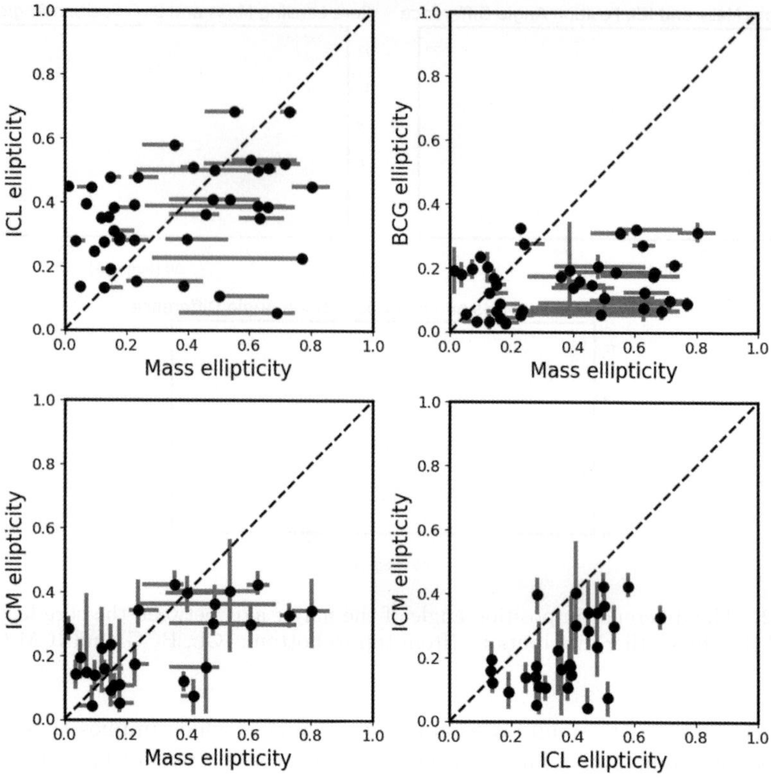

Figure 3. The ellipticity comparisons of various distributions (top-left: core lensing mass and ICL, top-right: core lensing mass and BCG, bottom-left: core lensing mass and ICM, bottom-right: ICL and ICM)

calculating the difference in centroid, we use the difference in projected radius as defined by $\Delta R = |r_1 - r_2|$ in units of kpc.

As expected, we measure small deviations in centroid for the ICL and BCG when compared to the core lensing mass. This implies that generally the BCG and ICL lie close to the center of the dark matter gravitational potential and that typically their mutual alignments coincide. The ICL is slightly more aligned with the core lensing mass distribution centroid than the BCG. This is because the BCG experiences a greater residual wobble around the dark matter gravitational center than the ICL.

Even after excluding double cored systems undergoing obvious major mergers, we still see that the ICM Gas is displaced to a much larger projected radius than the ICL or BCG. This is likely due to the fact that the ICM Gas "sloshes" as a consequence of major merger activity in the cluster's past (Markevitch et al. 2001; Churazov et al. 2003; Johnson et al. 2012; Harvey et al. 2017). The sloshing behavior is found to persist in cases where the BCG and ICL distributions have relaxed back to the core lensing mass centroid since these oscillations are damped on a much longer time scale due to hydrodynamical processes that only affect the ICM (Markevitch and Vikhlinin 2007).

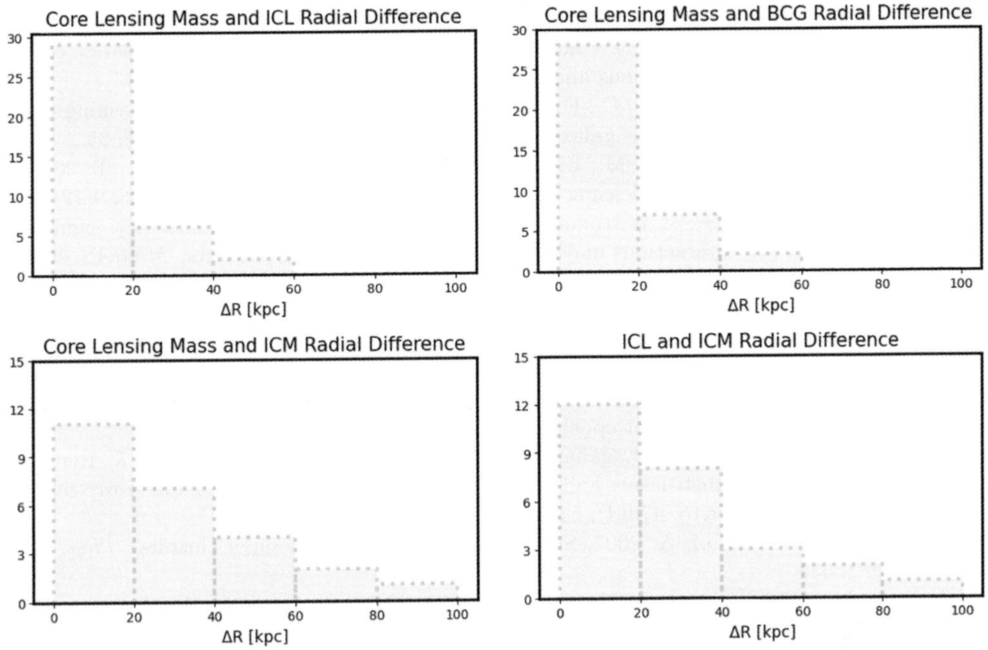

Figure 4. The centroid comparisons of various distributions (top-left: core lensing mass and ICL, top-right: core lensing mass and BCG, bottom-left: core lensing mass and ICM, bottom-right: ICL and ICM). We exclude doubly cored systems with extrememly large deviations

References

Arena, S. E. & Bertin, G. 2007, Slow evolution of elliptical galaxies induced by dynamical friction. III. Role of density concentration and pressure anisotropy. *AAP*, 463(3), 921–935.

Beers, T. & Geller, M. 1983, The environment of d and cd galaxies. *ApJ*, 274, 491.

Biernacka, M., Panko, E., Bajan, K., Godłowski, W., & Flin, P. 2015, The alignment of galaxy structures. *ApJ*, 813(1), 20.

Binggeli, B. 1982, The shape and orientation of clusters of galaxies. *PASP*, 107, 338.

Churazov, E., Forman, W., Jones, C., & Böhringer, H. 2003, Xmm-newton observations of the perseus cluster. i. the temperature and surface brightness structure. *ApJ*, 590(1), 225.

Contini, E., Chen, H. Z., & Gu, Q. 2022, The transition region between brightest cluster galaxies and intracluster light in galaxy groups and clusters. *ApJ*, 928, 99.

De Propris, R., West, M. J., Andrade-Santos, F., Ragone-Figueroa, C., Rasia, E., Forman, W., Jones, C., Kipper, R., Borgani, S., Lambas, D. G., Romashkova, E. A., & Patra, K. C. 2021, Brightest cluster galaxies: the centre can(not?) hold. *MNRAS*, 500(1), 310–318.

Donahue, M., Ettori, S., Rasia, E., Sayers, J., Zitrin, A., Meneghetti, M., Voit, G. M., Golwala, S., Czakon, N., Yepes, G., Baldi, A., Koekemoer, A., & Postman, M. 2016, The morphologies and alignments of gas, mass, and the central galaxies of clash clusters of galaxies. *ApJ*, 819(1), 36.

Harvey, D., Courbin, F., Kneib, J. P., & McCarthy, I. G. 2017, A detection of wobbling brightest cluster galaxies within massive galaxy clusters. *MNRAS*, 472(2), 1972–1980.

Hashimoto, Y., Henry, J. P., & Boehringer, H. 2008, Alignment of galaxies and clusters. *MNRAS*, 390(4), 1562–1568.

Hashimoto, Y., Henry, J. P., & Boehringer, H. 2014, Multiwavelength investigations of co-evolution of bright cluster galaxies and their host clusters. *MNRAS*, 440(1), 588–600.

Hikage, C., Mandelbaum, R., Takada, M., & Spergel, D. N. 2013, Where are the Luminous Red Galaxies (LRGs)? Using correlation measurements and lensing to relate LRGs to dark matter haloes. *MNRAS*, 435(3), 2345–2370.

Hoshino, H., Leauthaud, A., Lackner, C., Hikage, C., Rozo, E., Rykoff, E., Mandelbaum, R., More, S., More, A., Saito, S., & Vulcani, B. 2015, Luminous red galaxies in clusters: central occupation, spatial distributions and miscentring. *MNRAS*, 452(1), 998–1013.

Johnson, R. E., Zuhone, J., Jones, C., Forman, W. R., & Markevitch, M. 2012, Sloshing gas in the core of the most luminous galaxy cluster rxj1347.5-1145. *MNRAS*, 751(2), 95.

Jullo, E., Kneib, J.-P., Limousin, M., Elíasdóttir, Á., Marshall, P. J., & Verdugo, T. 2007, A bayesian approach to strong lensing modelling of galaxy clusters. *New J. Phys.*, 9(12), 447.

Kim, S. Y., Peter, A. H. G., & Wittman, D. 2017, In the wake of dark giants: new signatures of dark matter self-interactions in equal-mass mergers of galaxy clusters. *MNRAS*, 469(2), 1414–1444.

Lange, J. U., van den Bosch, F. C., Hearin, A., Campbell, D., Zentner, A. R., Villarreal, A. S., & Mao, Y.-Y. 2018, Brightest galaxies as halo centre tracers in SDSS DR7. *MNRAS*, 473(2), 2830–2851.

Lauer, T. R., Postman, M., Strauss, M. A., Graves, G. J., & Chisari, N. E. 2014, Brightest cluster galaxies at the present epoch. *ApJ*, 797(2), 82.

Lopes, P. A. A., Trevisan, M., Laganá, T. F., Durret, F., Ribeiro, A. L. B., & Rembold, S. B. 2018, Optical substructure and BCG offsets of Sunyaev–Zel'dovich and X-ray-selected galaxy clusters. *MNRAS*, 478(4), 5473–5490.

Markevitch, M. & Vikhlinin, A. 2007, Shocks and cold fronts in galaxy clusters. *Phys. Rep.*, 443(1), 1–53.

Markevitch, M., Vikhlinin, A., & Mazzotta, P. 2001, Nonhydrostatic gas in the core of the relaxed galaxy cluster a1795. *ApJ*, 562(2), L153.

Martel, H., Robichaud, F., & Barai, P. 2014, Major cluster mergers and the location of the brightest cluster galaxy. *ApJ*, 786(2), 79.

Montes, M. & Trujillo, I. 2018, Intracluster light: a luminous tracer for dark matter in clusters of galaxies. *MNRAS*, 482(2), 2838–2851.

Niederste-Ostholt, M., Strauss, M. A., Dong, F., Koester, B. P., & McKay, T. A. 2010, Alignment of brightest cluster galaxies with their host clusters. *MNRAS*, 405(3), 2023–2036.

Oliva-Altamirano, P., Brough, S., Lidman, C., Couch, W. J., Hopkins, A. M., Colless, M., Taylor, E., Robotham, A. S. G., Gunawardhana, M. L. P., Ponman, T., Baldry, I., Bauer, A. E., Bland-Hawthorn, J., Cluver, M., Cameron, E., Conselice, C. J., Driver, S., Edge, A. C., Graham, A. W., van Kampen, E., Lara-López, M. A., Liske, J., López-Sánchez, A. R., Loveday, J., Mahajan, S., Peacock, J., Phillipps, S., Pimbblet, K. A., & Sharp, R. G. 2014, Galaxy And Mass Assembly (GAMA): testing galaxy formation models through the most massive galaxies in the Universe. *MNRAS*, 440(1), 762–775.

Rossetti, M., Gastaldello, F., Ferioli, G., Bersanelli, M., De Grandi, S., Eckert, D., Ghizzardi, S., Maino, D., & Molendi, S. 2016, Measuring the dynamical state of Planck SZ-selected clusters: X-ray peak – BCG offset. *MNRAS*, 457(4), 4515–4524.

Sanderson, A. J. R., Edge, A. C., & Smith, G. P. 2009, LoCuSS: the connection between brightest cluster galaxy activity, gas cooling and dynamical disturbance of X-ray cluster cores. *MNRAS*, 398(4), 1698–1705.

Sastry, G. N. 1968, Clusters associated with supergiant galaxies. *PASP*, 80(474), 252.

Sharon, K., Bayliss, M. B., Dahle, H., Dunham, S. J., Florian, M. K., Gladders, M. D., Johnson, T. L., Mahler, G., Paterno-Mahler, R., Rigby, J. R., Whitaker, K. E., Akhshik, M., Koester, B. P., Murray, K., González, J. D. R., & Wuyts, E. 2020, Strong lens models for 37 clusters of galaxies from the sdss giant arcs survey*. *ApJS*, 247(1), 12.

Sharon, K., Cerny, C., Rigby, J. R., Florian, M. K., Bayliss, M. B., Dahle, H., Gladders, M. D., & Mahler, G. 2022,a HST-Based Lens Model of SDSS J1226+2152, in Preparation for JWST-ERS TEMPLATES. *arXiv e-prints*, arXiv:2207.05709.

Sharon, K., Mahler, G., Rivera-Thorsen, T. E., Dahle, H., Gladders, M. D., Bayliss, M. B., Florian, M. K., Kim, K. J., Khullar, G., Mainali, R., Napier, K. A., Navarre, A., Rigby, J. R., Remolina González, J. D., & Sharma, S. 2022,b The Cosmic Telescope That Lenses the Sunburst Arc, PSZ1 G311.65-18.48: Strong Gravitational Lensing Model and Source Plane Analysis. *ApJ*, 941b(2), 203.

Skibba, R. A., van den Bosch, F. C., Yang, X., More, S., Mo, H., & Fontanot, F. 2011, Are brightest halo galaxies central galaxies? *MNRAS*, 410(1), 417–431.

Van Den Bosch, F. C., Weinmann, S. M., Yang, X., Mo, H. J., Li, C., & Jing, Y. P. 2005, The phase-space parameters of the brightest halo galaxies. *MNRAS*, 361(4), 1203–1215.

Wang, L., Yang, X., Shen, S., Mo, H. J., van den Bosch, F. C., Luo, W., Wang, Y., Lau, E. T., Wang, Q. D., Kang, X., & Li, R. 2014, Measuring the X-ray luminosities of SDSS DR7 clusters from ROSAT All Sky Survey. *MNRAS*, 439(1), 611–622.

Wang, P., Luo, Y., Kang, X., Libeskind, N. I., Wang, L., Zhang, Y., Tempel, E., & Guo, Q. 2018, Alignment between satellite and central galaxies in the sdss dr7: Dependence on large-scale environment. *ApJ*, 859(2), 115.

Wittman, D., Foote, D., & Golovich, N. 2019, Brightest cluster galaxy alignments in merging clusters. *ApJ*, 874(1), 84.

Zenteno, A., Hernández-Lang, D., Klein, M., Vergara Cervantes, C., Hollowood, D. L., Bhargava, S., Palmese, A., Strazzullo, V., Romer, A. K., Mohr, J. J., Jeltema, T., Saro, A., Lidman, C., Gruen, D., Ojeda, V., Katzenberger, A., Aguena, M., Allam, S., Avila, S., Bayliss, M., Bertin, E., Brooks, D., Buckley-Geer, E., Burke, D. L., Capasso, R., Carnero Rosell, A., Carrasco Kind, M., Carretero, J., Castander, F. J., Costanzi, M., da Costa, L. N., De Vicente, J., Desai, S., Diehl, H. T., Doel, P., Eifler, T. F., Evrard, A. E., Flaugher, B., Floyd, B., Fosalba, P., Frieman, J., García-Bellido, J., Gerdes, D. W., Gonzalez, J. R., Gruendl, R. A., Gschwend, J., Gutierrez, G., Hartley, W. G., Hinton, S. R., Honscheid, K., James, D. J., Kuehn, K., Lahav, O., Lima, M., McDonald, M., Maia, M. A. G., March, M., Melchior, P., Menanteau, F., Miquel, R., Ogando, R. L. C., Paz-Chinchón, F., Plazas, A. A., Roodman, A., Rykoff, E. S., Sanchez, E., Scarpine, V., Schubnell, M., Serrano, S., Sevilla-Noarbe, I., Smith, M., Soares-Santos, M., Suchyta, E., Swanson, M. E. C., Tarle, G., Thomas, D., Varga, T. N., Walker, A. R., Wilkinson, R. D., & Collaboration), D. 2020, A joint SZ–X-ray–optical analysis of the dynamical state of 288 massive galaxy clusters. *MNRAS*, 495(1), 705–725.

Zitrin, A., Bartelmann, M., Umetsu, K., Oguri, M., & Broadhurst, T. 2012, Miscentring in galaxy clusters: dark matter to brightest cluster galaxy offsets in 10000 Sloan Digital Sky Survey clusters. *MNRAS*, 426(4), 2944–2956.

Strong Gravitational Lensing in the Era of Big Data
Proceedings IAU Symposium No. 381, 2024
H. Stacey, A. Sonnenfeld & C. Grillo, eds.
doi:10.1017/S1743921323004015

MAximum-entropy ReconStruction (MARS): A New Strong-lensing Reconstruction Algorithm for the JWST Era

Sangjun Cha[1] and M. James Jee[1,2]

[1]Department of Astronomy, Yonsei University, 50 Yonsei-ro, Seoul 03722, Korea.
email: sang6199@yonsei.ac.kr

[2]Department of Physics and Astronomy, University of California, Davis, One Shields Avenue, Davis, CA 95616, USA

Abstract. The MAximum-entropy ReconStruction (MARS) method is a free-form strong-lensing (SL) reconstruction algorithm, which adopts the maximum cross-entropy as a regularization. MARS shows remarkable convergence of multiple images in both source ($\sim 0.''02$) and image planes ($\sim 0.''05 - 0.''1$) while suppressing spurious fluctuations. Although the reconstruction requires a large number of free parameters exceeding ∼19,000, our implementation through PyTorch can obtain the reconstruction within hours. From our test using the publicly available synthetic clusters, we have verified that the reconstructed radial mass profiles are consistent with the truth within 1 percent. This makes MARS one of the best-performing SL reconstruction methods. We apply MARS to the six Hubble Frontier Fields clusters and present new mass reconstruction results. We also reconstruct a mass model of Abell 2744 using both weak-lensing (WL) and SL data from the JWST observations, with the largest dataset of Abell 2744, including 286 SL multiple images and ~ 350 arcmin^{-2} WL constraints.

Keywords. Dark matter distribution, Galaxy clusters, Weak gravitational lensing, Strong gravitational lensing

1. Introduction

Thanks to its advantage of not requiring any dynamical assumptions, gravitational lensing is one of the most powerful methods for reconstructing the mass distributions of astronomical objects (Bartelmann & Schneider 2001). Gravitational lensing is usually classified into two types: strong lensing (SL) and weak lensing (WL). WL uses the shape distortion of background galaxies and can provide mass maps for a wide area. SL utilizes the locations of multiply-lensed galaxies and enables us to reconstruct the precise mass distribution around cluster centers (Kneib & Natarajan 2011).

In this study, we reconstruct SL lens models for six galaxy clusters in the Hubble Frontier Fields (HFF) programs by employing a free-form lens modeling algorithm, MARS, which adopts the maximum cross-entropy as a regularization (Cha & Jee 2022, 2023). MARS has shown one of the best performances among various lens modeling algorithms (Cha & Jee 2022), evaluated using the synthetic galaxy clusters Ares and Hera (Meneghetti et al. 2017). For the galaxy cluster Abell 2744 (A2744), one of the targets of the HFF program, we simultaneously reconstruct a mass model using WL and SL data from the JWST observations, including 286 SL multiple images and $\sim$ 350 arcmin^{-2} WL constraints. This dataset is the largest ever used for cluster mass reconstruction.

2. Data

For the SL modeling of the HFF clusters, we adopt the multiple image catalogs in Cha & Jee (2023), which compile multiple images from the literature. These multiple images are classified into two classes: gold and silver. Images with spectroscopic redshifts are considered gold images, while those with only photometric redshifts are categorized as silver images. Additionally, images agreed to be lensed by multiple previous studies are also considered silver images. We treat their redshifts as free parameters for the mass reconstruction process. Each cluster has 100 – 200 multiple images, and all of them are converged in both source and lens planes.

By utilizing the publicly available NIRCam mosaic images processed by the UNCOVER team† (Bezanson et al. 2022; Weaver et al. 2023), we can reconstruct the mass distribution of A2744 over a wider field than that covered by the HFF program (~ 2 Mpc $\times$ 2 Mpc region). We adopt the SL catalog from Cha & Jee (2023) for the region where the HFF data covers. For other regions, we compiled multiple images from Furtak et al. (2023) and Bergamini et al. (2023). Moreover, we identified 16 new multiple-image candidates from 6 systems. WL background sources are extracted based on photometric redshifts provided by the UNCOVER team. We measure the shapes of the sources using the F200W filter from JWST. The final dataset consists of 286 SL multiple images, and the WL source density reaches ~ 350 arcmin^{-2}, which is the largest dataset ever used for cluster mass reconstruction.

3. Result

Figure 1 displays the reconstructed mass maps of the HFF clusters. Although the `MARS` algorithm is entirely blind to the locations of the cluster galaxies, the reconstructed mass maps are highly consistent with their locations, while producing small scatters of the multiple images in both the source ($\sim 0.''02$) and the image ($\sim 0.''1$) planes. The lens plane scatters from `MARS` are lower than the previous results by a factor of 5-10 (see Figure 2). The most impressive case is MACS J0416.1-2403, which has 236 multiple images, all with spectroscopic redshifts. `MARS` converges its multiple images with small scatters in the image plane ($\sim 0.''084$) while reconstructing a smooth mass map, effectively suppressing spurious fluctuations. For more details, we refer readers to Cha & Jee (2023).

Figure 3 shows the reconstructed mass distribution of A2744 using both WL and SL datasets from JWST. Similar to the HFF clusters, `MARS` successfully traces the brightest cluster galaxies (BCGs) with the five strongest mass peaks. Not only for the BCGs, but the resulting mass map also shows good agreement with the number and luminosity density of the cluster member galaxies. Furthermore, the mass elongations in the outskirts align consistently with the directions reported for the large-scale filamentary structures observed in the X-ray hot gas (Eckert et al. 2015).

4. Conclusion

We have presented new mass maps of the HFF clusters using the `MARS` algorithm. Especially for A2744, we ran the mass modeling using both WL and SL datasets from JWST, which are the largest datasets ever used for cluster mass reconstruction. Although `MARS` is entirely blind to the locations of cluster member galaxies, the resulting mass model successfully traces the locations of BCGs. Moreover, even with more than ~19,000 free parameters to minimize, the `MARS` models exhibit excellent convergence in both source and image planes while effectively suppressing spurious fluctuations. By incorporating the WL dataset for mass reconstruction, `MARS` can predict the mass distribution not only

† `https://jwst-uncover.github.io/DR1.html`

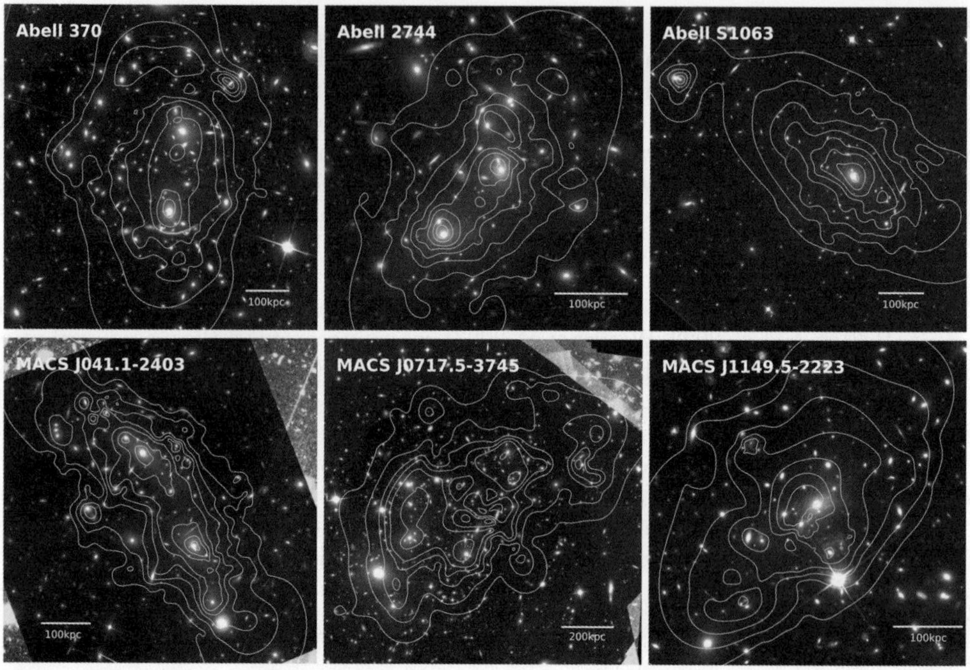

Figure 1. Mass contours of the HFF clusters overlaid on the color composite images. The white contours indicate the convergence κ. The color composite images are generated with the F606W, F814W, and F105W filters for the blue, green, and red channels, respectively except for Abell S1063. The color composite image of Abell S1063 is created with the F435W, F606W, and F814W filters for the blue, green, and red channels, respectively.

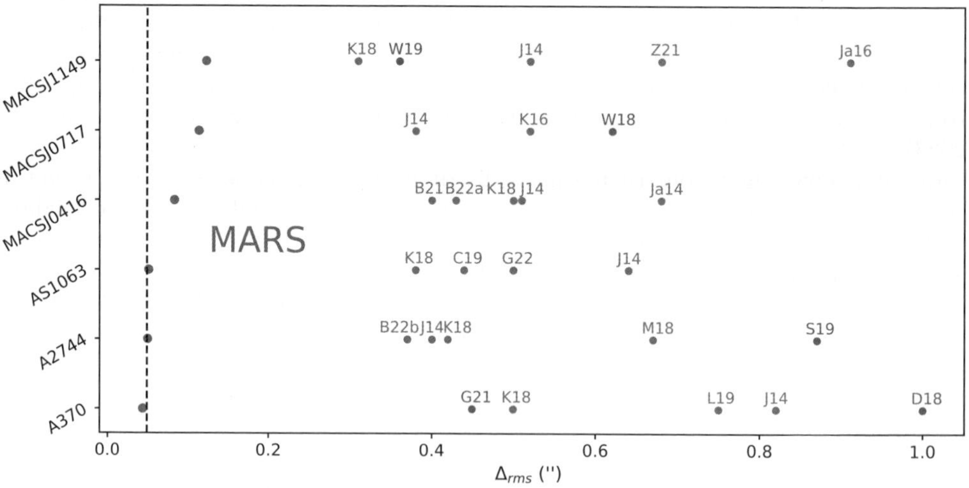

Figure 2. Image plane scatter comparison between `MARS` and the literature. The green (blue) dots indicate the RMS from the parametric (free-form) methods. The red dots show the RMS from `MARS`. The black dashed line represents an RMS value of 0.″05.

in the region around BCGs but also in the outskirts where there are no SL multiple images. In the current JWST era, with forthcoming observations expected to increase the dataset volumes of both SL and WL, a highly flexible and solidly regularized free-form modeling method like `MARS` would become a powerful tool to figure out precise mass

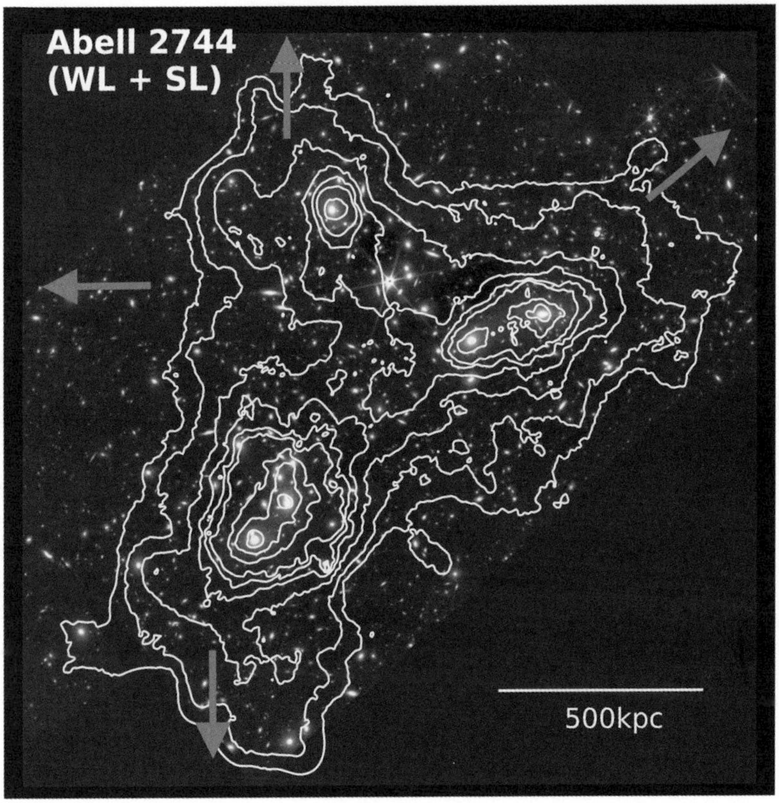

Figure 3. Reconstructed mass contours of A2744 overlaid on the color-composite image. The yellow contours are derived from the mass map using both WL and SL datasets. The red arrows indicate the reported directions of extended filaments observed in the X-ray hot gas. The color composite image is created using the F444W filter for red, the F277W filter for green, and the F115W filter for blue. The FOV is $400'' \times 400''$.

distributions of galaxy clusters across wide fields, from the central region to the outskirts and even beyond the virial radius.

Supplementary material

To view supplementary material for this article, please visit http://dx.doi.org/10.1017/S1743921323004015.

References

Bartelmann, M., & Schneider, P. 2001, *PhR*, 340, 291
Bergamini, P., Acebron, A., Grillo, E., *et al.* 2023, *ApJ*, 952, 84
Bezanson, R., Labbe, I., Whitaker, K. E., *et al.* 2022, arXiv:2212.04026
Cha, S., & Jee, M. J. 2022, *ApJ*, 931, 127
Cha, S., & Jee, M. J. 2023, *ApJ*, 951, 140
Eckert, D., Jauzac, M., Shan, H., *et al.* 2015, *Nature*, 528, 105
Furtak, L. J., Zitrin, A., Weaver, J. R., *et al.* 2023, *MNRAS*, 523, 4568
Kenib, J. -P., & Natarajan, P. 2011, *A&ARv*, 19, 47
Meneghetti, M., Natarajan, P., Coe, D., *et al.* 2017, *MNRAS*, 483, 195
Weaver, J. R., Cutler, S. E., Pan, R., *et al.* 2023, arXiv:2301.02671

Chapter 4

Strong Gravitational Lensing in the Era of Big Data
Proceedings IAU Symposium No. 381, 2024
H. Stacey, A. Sonnenfeld & C. Grillo, eds.
doi:10.1017/S1743921323003964

The present & future of lensed supernovae: from ZTF to LSST

Nikki Arendse

Oskar Klein Centre, Department of Physics, Stockholm University, SE-106 91 Stockholm, Sweden. email: nikki.arendse@fysik.su.se

Abstract. Gravitationally lensed supernovae (SNe) are rare and valuable probes of astrophysics and cosmology. While only seven lensed SNe have currently been discovered, these numbers are predicted to increase by orders of magnitudes with future transient surveys such as the Legacy Survey of Space and Time (LSST). These proceedings describe the ongoing live search with the Zwicky Transient Facility (ZTF), including the discovery of SN 'Zwicky': a lensed SN found in a remarkably low-mass lens galaxy. Finally, we look ahead at predictions for detecting lensed SNe with LSST.

Keywords. Strong gravitational lensing, supernovae, cosmology

1. Introduction

Strongly lensed supernovae (SNe) can give valuable insights into the expansion rate of the Universe, substructures in lens galaxies, and high-redshift supernova physics. When the time delay between the appearance of the multiple SN images is combined with a model of the gravitational potential of the lens galaxy, a measurement of the Hubble constant (H_0) can be obtained which is independent of the distance ladder and Cosmic Microwave Background (CMB) radiation. Currently, five strongly lensed SNe have been discovered in galaxy clusters: SN Refsdal (Kelly *et al.* 2015), SN Requiem (Rodney *et al.* 2021), 2022riv (Kelly *et al.* 2022), C22 (Chen *et al.* 2022), and SN H0pe (Frye *et al.* 2023). Additionally, two type Ia SNe have been found that were lensed by single elliptical galaxies: iPTF16geu (Goobar *et al.* 2017) and SN Zwicky (Goobar *et al.* 2023). For the cluster-scale lensed SNe, the limiting factor for precision cosmology is the galaxy cluster's complicated mass model. The two galaxy-scale lensed SNe, on the other hand, have a relatively simple mass model but extremely short time delays. In these conference proceedings, we describe the ongoing lensed SN search with the Zwicky Transient Facility (ZTF) and look ahead at forecasts for the upcoming Legacy Survey of Space and Time (LSST) to be conducted at the Vera Rubin Observatory.

2. Discovery of lensed SN 'Zwicky'

The ZTF, located in Palomar, California, has been observing the northern sky every 2 to 3 days for the past four years. Besides discovering SNe and other transients routinely, it has the potential to observe a handful of lensed SNe in its lifetime. On August 1st 2022, it detected a supernova (SN 2022qmx, also known as 'SN Zwicky') that was bright enough to pass the threshold of automatic spectroscopic follow-up. The resulting spectrum, taken by SED Machine, revealed a type Ia supernova (SNIa) at a redshift of $z = 0.35$ that was ~ 25 times as bright as standard type Ia SNe. When the supernova redshift was confirmed by a spectrum from the Nordic Optical Telescope (NOT), our team

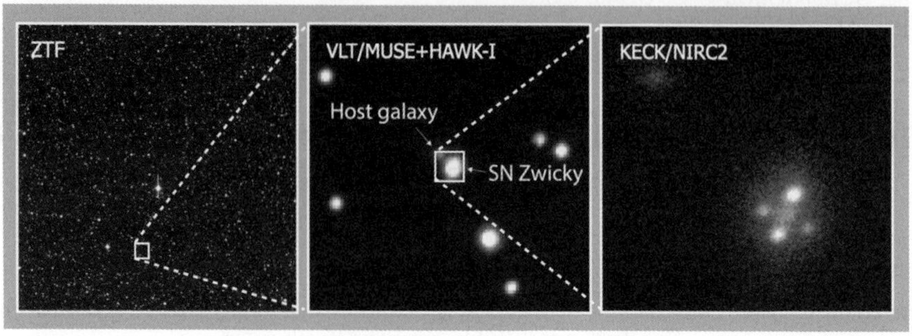

Figure 1. Zooming into lensed SN Zwicky: from ZTF to VLT to Keck observations. Adaptive Optics with Keck finally managed to resolve the four images of SN Zwicky. The host galaxy of the SN is located in the upper left corner.

knew that the supernova had to be gravitationally magnified to explain its exceptional brightness. We conducted follow-up observations with the Very Large Telescope (VLT) in Chile, which provided us with high-resolution spectra to improve the redshift estimates. However, VLT observations were unable to resolve the multiple images. The four images of SN Zwicky only became visible with observations from the Keck telescope in Hawaii using adaptive optics, as displayed in Figure 1. The Keck data allowed us to construct a lens model from the image positions and infer an Einstein radius of only $\theta_E = 0.167''$, one of the most compact lens systems ever discovered in the optical. A few days later, SN Zwicky was also observed with the Hubble Space Telescope (HST), which was also able to resolve the lens system (Pierel *et al.* 2023). SN Zwicky and the previously discovered galaxy-scale lensed SNIa, iPTF16geu, form a new population of low-mass and small image separation lens systems, as illustrated in Figure 2 (Goobar *et al.* 2023). Standard candles such as SNIa are a unique tool to discover and characterize the properties of these light, compact galaxy-scale gravitational lenses. However, this new population of lens galaxies are highly magnified, symmetric systems which often have very short time delays. The best-fit maximum time delay of SN Zwicky, obtained from the unresolved light curve and Keck flux ratios, is less than a day. Although such short time delays make these low-mass lens systems inconvenient for time-delay cosmography, they can reveal a lot about lens galaxy substructures and high-redshift supernova physics. The observed flux ratios of SN Zwicky are discrepant with the results from our best-fit lens model, hinting at additional magnification from substructures or microlensing of stars in the lens galaxy.

3. Prospects for detecting lensed SNe with LSST

Although the current sample of lensed SN discoveries is small, these numbers are about to increase by a few orders of magnitude with the advance of the next generation of survey telescopes such as the Vera Rubin Observatory and the Nancy Grace Roman Space Telescope. In this section, we focus on predictions for lensed SNIa discoveries in LSST. This survey will take multi-colour $ugrizy$ images and cover $\sim$20,000 square degrees of the sky in a ten-year period. 90% of observing time will be spent on the Wide-Fast-Deep (WFD) survey, consisting of an area of 18,000 square degrees. The WFD survey is expected to proceed using a rolling cadence, in which certain areas of the footprint will be assigned more frequent visits than others, in order to improve the light curve sampling of objects in the high-cadence regions.

In order to assess the impact of the observing strategy on lensed SNIa detections, we simulate a large data set of lensed SNIa light curves. We include microlensing effects and take cadence simulations from the OpSim database. We consider two strategies for

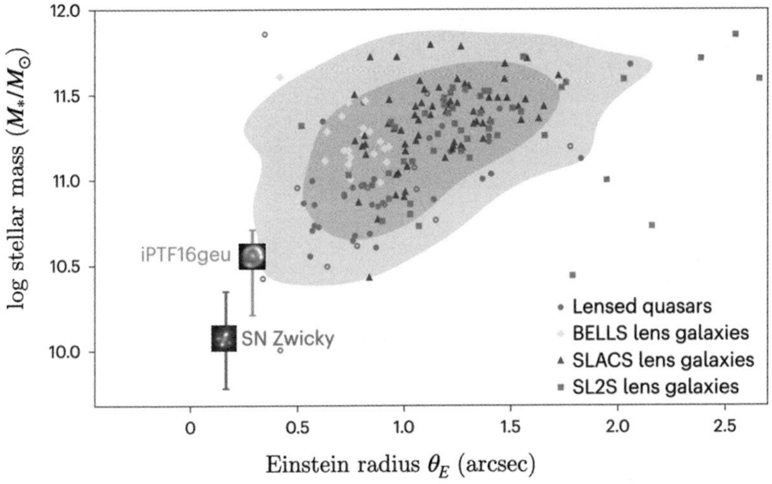

Figure 2. A comparison of SN Zwicky and iPTF16geu with other galaxy-scale lenses. Both in terms of stellar mass and Einstein radius, Zwicky and iPTF16geu are outliers. They point to a new population of low-mass and small-image-separation lensing systems.

detecting lensed SNe: the image multiplicity method, which looks for multiple resolved images, and the magnification method, which searches for objects that appear significantly brighter than a typical SNIa at the redshift of the lens galaxy (Wojtak *et al.* 2019). As a result, we predict to find ~ 44 lensed SNIa in the WFD survey per year, of which ~ 26 have time delays longer than 10 days (Arendse *et al.* in prep.). The effects of microlensing lower the predicted lensed SNIa detections by $\sim 8\%$. We also investigated how to separate the populations of lensed and unlensed SNIa. Since LSST will also discover unlensed SNIa at high redshifts (up to $z \sim 1$), looking for objects that appear 'redder' will not be sufficient anymore. Instead, we should combine colours with apparent peak magnitudes, since lensed SNIa are magnified and hence brighter than unlensed ones at the same redshifts (Arendse *et al.* in prep.).

4. Conclusion

We are entering an exciting era where the number of lensed SN discoveries will increase by several orders of magnitude with the advance of LSST and other upcoming surveys. Presently, we have found two galaxy-scale lensed SNIa, SN Zwicky and iPTF16geu. These are both symmetric, high-magnification lens systems with short time delays. They are part of a new population of very low-mass and compact lens galaxies, and are valuable for learning more about lens galaxy substructures and high-redshift supernova physics. With the upcoming Vera Rubin Observatory, we are expecting to discover around 40 lensed SNIa per year. However, many challenges remain that relate to the identification of these rare objects. We found that a decent separation between lensed and unlensed SNIa can be achieved when considering jointly their colours and apparent peak magnitudes.

References

Kelly, P. L., *et al.* 2015, *Science*, 347, 6226, 1123–1126

Rodney, S. A., *et al.* 2021, *Nature Astronomy*, 5, 1118–1125

Kelly, P. L., *et al.*, 2022, *Transient Name Server AstroNote*, 169, 1

Chen, W., *et al.*, 2022, *Nature*, 611, 7935, 256–259

Frye, B., *et al.*, 2023, *Transient Name Server AstroNote*, 96, 1

Goobar, A. M., 2017, *Science*, 356, 6335, 291–295

Goobar, A. M., 2023, *Nature Astronomy*
Pierel, J. D. R., 2023, *ApJ*
Wojtak, R. J., 2019, *MNRAS*, 487, 3, 3342–3355.
Arendse, N., *in prep.*

Strong Gravitational Lensing in the Era of Big Data
Proceedings IAU Symposium No. 381, 2024
H. Stacey, A. Sonnenfeld & C. Grillo, eds.
doi:10.1017/S1743921323003678

Extended surface brightness modeling of three sources strongly lensed by an ultra-massive elliptical galaxy

Andrea Bolamperti

Dipartimento di Fisica e Astronomia, Università degli Studi di Padova, Vicolo dell'Osservatorio 3, I-35122 Padova, Italy
Istituto Nazionale di Astrofisica (INAF), Osservatorio di Padova, Vicolo dell'Osservatorio 5, I-35122 Padova, Italy
European Southern Observatory, Karl-Schwarzschild-Strasse 2, D-85748 Garching bei München, Germany. email: andrea.bolamperti@phd.unipd.it

Abstract. Rare systems with multiple sources strongly lensed by a single galaxy provide the most robust way to explore the geometry of the Universe and to study the inner mass structure of lens galaxies. We present here a study of the SDSS J0100+1818 deflector, analyzing its total and baryonic mass distributions. The system comprises an ultra-massive early-type galaxy, surrounded by ten multiple images of three background sources. Exploiting high-resolution HST photometry and VLT/X-shooter spectroscopy we conduct a strong lensing analysis with the software `GLEE` to reconstruct the complex surface brightness distributions of the background sources over approximately 7200 HST pixels. These results are presented in our recent paper, Bolamperti et al. (2023). Finally, we present some preliminary results from new VLT/MUSE observations, that will allow us to build a new strong lensing+dynamics joint model and measure the values of the total matter density and of the cosmological constant parameters, Ω_m and Ω_Λ.

Keywords. gravitational lensing: strong - galaxies: evolution - dark matter - cosmology

1. Multiple sources strongly lensed by the same galaxy-scale deflector

Strong gravitational lensing is a valuable tool for delving into galaxy evolution and exploring the properties of the Universe (e.g., Bartelmann 2010; Treu 2010). It offers precise measurements of total mass, including both luminous and dark matter components, with relative errors typically at the few percent level (e.g., Grillo 2010; Zitrin et al. 2012). Additionally, strong lensing facilitates the study of lensed background sources thanks to the magnification effect. By reconstructing their surface brightness distribution during the modeling process (e.g., Rizzo et al. 2021; Wang et al. 2022), or utilizing high magnification factors to investigate faint and small high-z galaxies, we gain valuable insights into the local feedback mechanisms driving the galaxy evolution (e.g., Förster Schreiber et al. 2009; Cañameras et al. 2017a; Cava et al. 2018).

The most massive gravitational lenses, like galaxy clusters, can generate dozens of multiple images from sources in a large redshift range, and have provided in the last decades crucial insights into the mass density slope of dark matter halos in their inner regions (Sand et al. 2004; Grillo et al. 2015; Bergamini et al. 2019), the measurement of the cosmological density parameter values (Jullo et al. 2010; Caminha et al. 2016; Grillo et al. 2020), and of the Hubble constant H_0 (Grillo et al. 2018). On the other hand, they have a complex total mass distribution which includes several mass components, that might introduce degeneracies between the model parameters. The study of galaxy-scale strong lenses allows for the reconstruction of the total and dark matter mass distributions, the determination of dark matter over total mass fraction (Gavazzi et al. 2007; Grillo et al. 2009; Suyu and Halkola 2010; Sonnenfeld et al. 2015; Schuldt et al. 2019), the inference of the stellar initial mass function (IMF) of the lens (e.g., Cañameras et al. 2017b; Barnabè et al. 2013; Sonnenfeld et al. 2019), and the detection of dark matter substructures (e.g., Vegetti et al. 2012; Hezaveh et al. 2016; Ritondale et al. 2019). Unfortunately, they do not usually present the features needed to pursue cosmological studies. In fact, it is only possible when kinematic data of the lenses are available (e.g., Grillo et al. 2008; Cao et al. 2012), when a variable source is lensed (e.g., a quasar, Suyu et al. 2013, 2017) and when several background sources are multiply imaged by the same lens galaxy, and the mass sheet degeneracy can be broken (Schneider 2014).

In this scenario, the discovery of ultra massive galaxies that act as lenses for two or more background sources plays a key role in measuring the values of cosmological parameters through strong lensing modeling. In fact, by considering the ratio of the two Einstein radii observable in a double-source system, one can constrain the value of Ω_m independently of that of H_0 (Collett et al. 2012). The presence of more than two multiply-imaged sources can further reduce the uncertainties on the total mass profile of the lens and on the cosmological parameters, but is extremely rare, with only a few known cases to date (Collett and Smith 2020; Tu et al. 2009; Tanaka et al. 2016; Schuldt et al. 2019; Wang et al. 2022).

We present here the results of the next unique case, SDSS J010049.18+181827.7 (hereafter, SDSS J0100+1818), presented in a recent paper by Bolamperti et al. (2023). This system is part of the Cambridge And Sloan Survey Of Wide ARcs in the skY (CASSOWARY) survey (Belokurov et al. 2009; Stark et al. 2013). The exceptionally large Einstein radius of ≈ 42 kpc, together with the results from our follow-up observations with the Nordic Optical Telescope (NOT), the Very Large Telescope (VLT) and the Hubble Space Telescope (HST) suggest that the SDSS J0100+1818 deflector is an uncommon strong lens and a candidate fossil system at intermediate redshift. Initially, only the background sources A and B, at the same redshift, were identified (see Figure 1), but now we recognize a total of six background sources. Finally, we report some preliminary results from new VLT/MUSE observations, that will allow us to build a new lensing+dynamics joint model and measure the values of the cosmological parameters Ω_m and Ω_Λ.

2. SDSS J010049.18+181827.7

Given its peculiar features, SDSS J0100+1818 benefits from a large collection of photometric and spectroscopic data from the most advanced facilities. It was firstly observed with the 2.5 m Nordic Optical Telescope (NOT; program 56-032, PI: R. Cañameras), to obtain color information for galaxies in the lens environment and to characterize their spectral energy distributions (SEDs). Later, it was observed with the HST Wide Field Camera 3 (WFC3) (program GO-15253; PI: R. Cañameras), in the F438W and F160W filters. HST observations reduced the PSF full width at half maximum (FWHM) by a factor ~ 10, allowing us to resolve AB into two different sources at the same redshift (see Figure 1). In order to measure the lens and source spectroscopic redshifts, and the lens stellar velocity dispersion, SDSS J0100+1818 was observed with VLT/X-shooter (program 091.A-0852, PI: L. Christensen) targeting the lens and the multiple images AB 1-3-4.

We measure the lens redshift $z = 0.581$, and a joint redshift of $z = 1.880$ for the two source components forming the image families A and B. We measure a stellar velocity dispersion value of the deflector of (451 ± 37) km s^{-1}, confirming the ultra massive nature of SDSS J0100+1818, that is among the rarest, most massive elliptical galaxies known at its redshift (Loeb and Peebles 2003). We also fit its surface brightness distribution

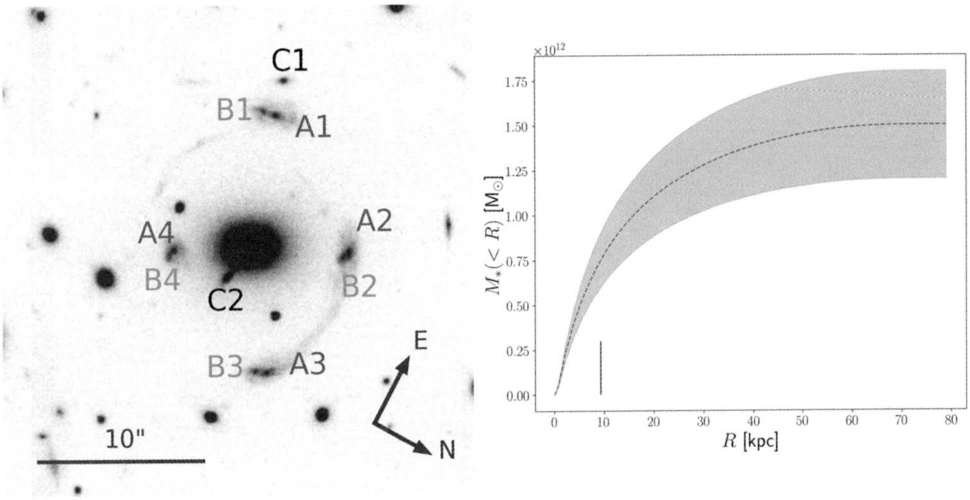

Figure 1. Left: scheme of the multiple images of SDSS J0100+1818, superimposed on the HST observations in the F160W filter. Multiple images of the same background source are labeled with the same letter. In Bolamperti et al. (2023), sources A and B are spectroscopically confirmed and included in the modeling, while C is included with its redshift as a free parameter. Right: cumulative stellar mass (M_*) profile and $\pm 1\sigma$ uncertainties of the main lens. It is measured from the total stellar mass measured through the SED fitting and assuming a constant stellar mass-to-light ratio. The blue vertical line shows the half-light radius of the lens galaxy.

and, by assuming a constant mass-to-light ratio, we estimate a total stellar mass value of $(1.5 \pm 0.3) \times 10^{12}$ M$_\odot$. Through a combination of photometric observations in the two available HST filters and preliminary strong lensing models of the system, we identify an additional candidate background source. In Figure 1, we label its two multiple images as C1 and C2. Although spectroscopic confirmation was unavailable, we incorporate source C into our strong lensing models, considering its redshift as a free parameter. This decision stems from the compelling evidence of two images having similar colors and being confidently predicted at the same positions by the strong lensing models. Given that C1 and C2 are, respectively, the furthest and closest images from the deflector, they play a key role in constraining the total mass distribution of the lens.

Considering the substantial angular separation between the multiple images, we explore the possibility of the main lens galaxy residing within a rich and overdense environment, as observed in other high-mass lens early-type galaxies at intermediate redshift (e.g., Newman et al. 2015; Johnson et al. 2018; Wang et al. 2022). We estimate the photometric redshifts of galaxies in a 2.5'×2.5' field of view, and our measurements broadly support the presence of an overdensity surrounding the main deflector. However, given the wide distribution of candidate members throughout the field, we opt to model its effect on the light deflection with an external shear component.

3. Strong lensing modeling

We employ the Gravitational Lens Efficient Explorer (`GLEE`; Suyu and Halkola 2010; Suyu et al. 2012) software to develop several strong lensing models of the deflector, combining different mass density profiles such as (cored or singular) pseudo-isothermal total mass distribution (PIEMD+rc, cored; Kneib et al. 1996) and singular power law elliptical mass distribution (SPEMD; Barkana 1998), with or without an external shear component. Initially, we treat the sources as point-like objects, exploiting the observed

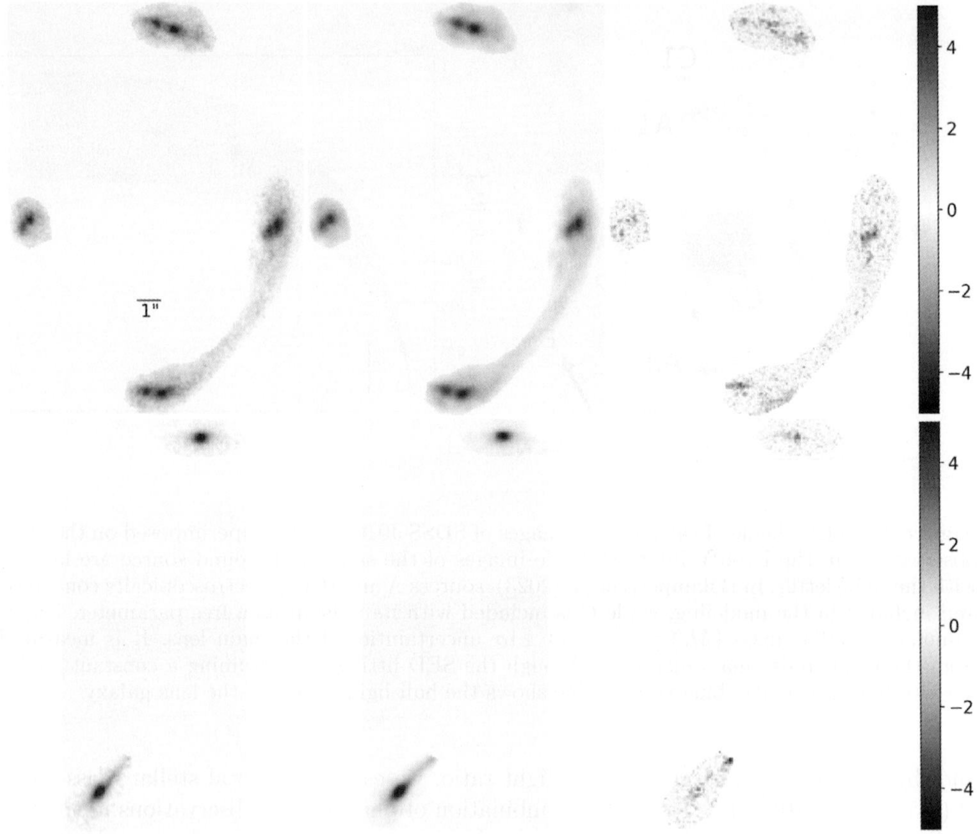

Figure 2. From the left to the right: observed SB in the F160W band of the multiple images considered in the extended strong lensing modeling; best-fit model-predicted SB, obtained by assuming a SPEMD model; normalized residuals in the range from -5σ to $+5\sigma$. The top row shows the sources A and B, while the bottom row shows the source C.

positions of the multiple images of source AB (four multiple images for each of the two components) and the two observed positions of the multiple images of source C.

Going beyond the point-like modeling, we consider the lensed sources as extended objects, enabling the reconstruction of their surface brightness (SB) distributions. This improvement allows us to leverage the detailed image structure and the extended arcs in which the sources were distorted, covering approximately 7200 HST pixels. The results are shown in Figure 2.

For each model, we estimate the total mass (M_T) distribution of the deflector. In the point-like source modeling, we obtain a projected total mass value within the Einstein radius (of approximately 42 kpc) of $(9.1 \pm 0.1) \times 10^{12}$ $M_\odot$, which remains consistent for both the PIEMD+rc and SPEMD models. The best-fit redshift estimate for source C is found to be $z_C = 1.72$ (1.98) for the PIEMD+rc (SPEMD) model. These measurements are consistent with the extended modeling. Considering the extended structure of the sources and their multiple images, the number of observables increases significantly to approximately 7200 (compared to the 20 observables in the point-like source modeling). As a result, the statistical uncertainties on the parameter values and derived quantities are drastically reduced (see Figure 3). This improvement in the modeling accuracy allows for a more precise characterization of the properties of the system and lensing effects on the background sources.

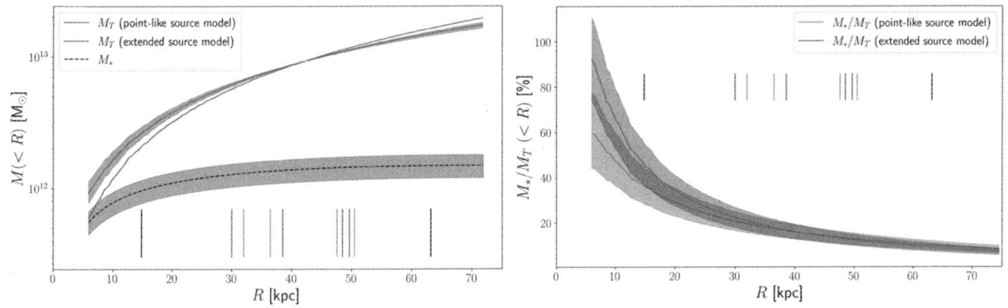

Figure 3. Left: cumulative projected total mass profiles for the SPEMD model, obtained from both point-like (blue) and extended source (red) modeling. The shaded areas represent the $\pm 1\sigma$ uncertainties, which are significantly reduced (smaller than the linewidth) for the extended source modeling. The vertical lines indicate the distances from the lens galaxy center of the various multiple images, color-coded as shown in Figure 1. Right: analogously to the left panel, but here we show the cumulative projected stellar-over-total mass fraction profiles.

4. VLT/MUSE observations

SDSS J0100+1818 has been targeted with VLT/MUSE (program 0110.A-0248, PI: A. Bolamperti) to obtain integral field spectroscopy of the system to identify other galaxies on the lens plane, spectroscopically confirm the redshift of system C, enhance our study on the total and dark-matter mass profiles, and measure the stellar velocity dispersion profile of the main deflector. In the 1'×1' MUSE field of view we find at least other 11 galaxies at $z \approx 0.58$, confirming that SDSS J0100+1818 is not an isolated galaxy, but it is more likely the brightest galaxy of a group residing in a dark matter halo. Furthermore, we confirm the stellar velocity dispersion measured with X-shooter, but we are now able to derive its radial profile. Most importantly, we spectroscopically confirm that C1 and C2 are two multiple images of the same background source and we identify, through their Lyman-α emission, two additional sources at higher redshift, each with four multiple images.

5. Discussion and conclusions

We have found SDSS J0100+1818 to be one of the few known galaxy-scale systems with several lensed sources at different redshifts (see e.g., Collett and Smith 2020). Such advancements hold great potential for refining our understanding of this complex gravitational lensing system. Our understanding of SDSS J0100+1818 will significantly benefit from additional integral field spectroscopy, offering several avenues for improvement:

(1) The presence of several background sources spanning a wide redshift range, and the simple mass distribution of the deflector, make SDSS J0100+1818 a unique system to probe the value of the cosmological parameters Ω_m and Ω_Λ.
(2) The inclusion of the redshift value of source C will help in breaking the degeneracy between the values of the Einstein radius and of the slope of the mass profile. Moreover, the redshift of C would enable a multi-plane lensing analysis, considering the mutual deflection of light between the different background sources.
(3) We can now include companions galaxies at the same redshift in the strong lensing model rather than treating them merely as an external shear component.
(4) By measuring the velocity dispersion profile of the main lens galaxy and combining the kinematic data with the strong lensing information, we could gain deeper insights into the dynamics of the system and its mass distribution.

References

Barkana, R. 1998, Fast Calculation of a Family of Elliptical Mass Gravitational Lens Models. *ApJ*, 502(2), 531–537.

Barnabè, M., Spiniello, C., Koopmans, L. V. E., *et al.* 2013, A low-mass cut-off near the hydrogen burning limit for Salpeter-like initial mass functions in early-type galaxies. *MNRAS*, 436(1), 253–258.

Bartelmann, M. 2010, TOPICAL REVIEW Gravitational lensing. *Classical and Quantum Gravity*, 27(23), 233001.

Belokurov, V., Evans, N. W., Hewett, P. C., *et al.* 2009, Two new large-separation gravitational lenses from SDSS. *MNRAS*, 392(1), 104–112.

Bergamini, P., Rosati, P., Mercurio, A., *et al.* 2019, Enhanced cluster lensing models with measured galaxy kinematics. *A&A*, 631, A130.

Bolamperti, A., Grillo, C., Cañameras, R., *et al.* 2023, Reconstructing the extended structure of multiple sources strongly lensed by the ultra-massive elliptical galaxy SDSS J0100+1818. *A&A*, 671, A60.

Cañameras, R., Nesvadba, N., Kneissl, R., *et al.* 2017,a Planck's dusty GEMS. IV. Star formation and feedback in a maximum starburst at z = 3 seen at 60-pc resolution. *A&A*, 604a, A117.

Cañameras, R., Nesvadba, N. P. H., Kneissl, R., *et al.* 2017,b Planck's dusty GEMS. III. A massive lensing galaxy with a bottom-heavy stellar initial mass function at z = 1.5. *A&A*, 600b, L3.

Caminha, G. B., Grillo, C., Rosati, P., *et al.* 2016, CLASH-VLT: A highly precise strong lensing model of the galaxy cluster RXC J2248.7-4431 (Abell S1063) and prospects for cosmography. *A&A*, 587, A80.

Cao, S., Pan, Y., Biesiada, M., *et al.* 2012, Constraints on cosmological models from strong gravitational lensing systems. *J. Cosmology Astropart. Phys.*, 2012(3), 016.

Cava, A., Schaerer, D., Richard, J., *et al.* 2018, The nature of giant clumps in distant galaxies probed by the anatomy of the cosmic snake. *Nature Astronomy*, 2, 76–82.

Collett, T. E., Auger, M. W., Belokurov, V., *et al.* 2012, Constraining the dark energy equation of state with double-source plane strong lenses. *MNRAS*, 424(4), 2864–2875.

Collett, T. E. & Smith, R. J. 2020, A triple rollover: a third multiply imaged source at z 6 behind the Jackpot gravitational lens. *MNRAS*, 497(2), 1654–1660.

Förster Schreiber, N. M., Genzel, R., Bouché, N., *et al.* 2009, The SINS Survey: SINFONI Integral Field Spectroscopy of z ~2 Star-forming Galaxies. *ApJ*, 706(2), 1364–1428.

Gavazzi, R., Treu, T., Rhodes, J. D., *et al.* 2007, The Sloan Lens ACS Survey. IV. The Mass Density Profile of Early-Type Galaxies out to 100 Effective Radii. *ApJ*, 667(1), 176–190.

Grillo, C. 2010, Projected Central Dark Matter Fractions and Densities in Massive Early-type Galaxies from the Sloan Digital Sky Survey. *ApJ*, 722(1), 779–787.

Grillo, C., Gobat, R., Lombardi, M., & Rosati, P. 2009, Photometric mass and mass decomposition in early-type lens galaxies. *A&A*, 501(2), 461–474.

Grillo, C., Lombardi, M., & Bertin, G. 2008, Cosmological parameters from strong gravitational lensing and stellar dynamics in elliptical galaxies. *A&A*, 477(2), 397–406.

Grillo, C., Rosati, P., Suyu, S. H., *et al.* 2018, Measuring the Value of the Hubble Constant "à la Refsdal". *ApJ*, 860(2), 94.

Grillo, C., Rosati, P., Suyu, S. H., *et al.* 2020, On the Accuracy of Time-delay Cosmography in the Frontier Fields Cluster MACS J1149.5+2223 with Supernova Refsdal. *ApJ*, 898(1), 87.

Grillo, C., Suyu, S. H., Rosati, P., *et al.* 2015, CLASH-VLT: Insights on the Mass Substructures in the Frontier Fields Cluster MACS J0416.1-2403 through Accurate Strong Lens Modeling. *ApJ*, 800(1), 38.

Hezaveh, Y. D., Dalal, N., Marrone, D. P., *et al.* 2016, Detection of Lensing Substructure Using ALMA Observations of the Dusty Galaxy SDP.81. *ApJ*, 823(1), 37.

Johnson, L. E., Irwin, J. A., White, Raymond E., I., *et al.* 2018, Using Strong Gravitational Lensing to Identify Fossil Group Progenitors. *ApJ*, 856(2), 131.

Jullo, E., Natarajan, P., Kneib, J. P., *et al.* 2010, Cosmological constraints from strong gravitational lensing in clusters of galaxies. *Science*, 329, 924–927.

Kneib, J. P., Ellis, R. S., Smail, I., *et al.* 1996, Hubble Space Telescope Observations of the Lensing Cluster Abell 2218. *ApJ*, 471, 643.

Loeb, A. & Peebles, P. J. E. 2003, Cosmological Origin of the Stellar Velocity Dispersions in Massive Early-Type Galaxies. *ApJ*, 589(1), 29–34.

Newman, A. B., Ellis, R. S., & Treu, T. 2015, Luminous and Dark Matter Profiles from Galaxies to Clusters: Bridging the Gap with Group-scale Lenses. *ApJ*, 814(1), 26.

Ritondale, E., Vegetti, S., Despali, G., *et al.* 2019, Low-mass halo perturbations in strong gravitational lenses at redshift z $\sim$ 0.5 are consistent with CDM. *MNRAS*, 485(2), 2179–2193.

Rizzo, F., Vegetti, S., Fraternali, F., *et al.* 2021, Dynamical properties of z 4.5 dusty star-forming galaxies and their connection with local early-type galaxies. *MNRAS*, 507(3), 3952–3984.

Sand, D. J., Treu, T., Smith, G. P., & Ellis, R. S. 2004, The Dark Matter Distribution in the Central Regions of Galaxy Clusters: Implications for Cold Dark Matter. *ApJ*, 604(1), 88–107.

Schneider, P. 2014, Can one determine cosmological parameters from multi-plane strong lens systems? *A&A*, 568, L2.

Schuldt, S., Chirivì, G., Suyu, S. H., *et al.* 2019, Inner dark matter distribution of the Cosmic Horseshoe (J1148+1930) with gravitational lensing and dynamics. *A&A*, 631, A40.

Sonnenfeld, A., Jaelani, A. T., Chan, J., *et al.* 2019, Survey of gravitationally-lensed objects in HSC imaging (SuGOHI). III. Statistical strong lensing constraints on the stellar IMF of CMASS galaxies. *A&A*, 630, A71.

Sonnenfeld, A., Treu, T., Marshall, P. J., *et al.* 2015, The SL2S Galaxy-scale Lens Sample. V. Dark Matter Halos and Stellar IMF of Massive Early-type Galaxies Out to Redshift 0.8. *ApJ*, 800(2), 94.

Stark, D. P., Auger, M., Belokurov, V., *et al.* 2013, The CASSOWARY spectroscopy survey: a new sample of gravitationally lensed galaxies in SDSS. *MNRAS*, 436(2), 1040–1056.

Suyu, S. H., Auger, M. W., Hilbert, S., *et al.* 2013, Two Accurate Time-delay Distances from Strong Lensing: Implications for Cosmology. *ApJ*, 766(2), 70.

Suyu, S. H., Bonvin, V., Courbin, F., *et al.* 2017, H0LiCOW - I. H_0 Lenses in COSMOGRAIL's Wellspring: program overview. *MNRAS*, 468(3), 2590–2604.

Suyu, S. H. & Halkola, A. 2010, The halos of satellite galaxies: the companion of the massive elliptical lens SL2S J08544-0121. *A&A*, 524, A94.

Suyu, S. H., Hensel, S. W., McKean, J. P., *et al.* 2012, Disentangling Baryons and Dark Matter in the Spiral Gravitational Lens B1933+503. *ApJ*, 750(1), 10.

Tanaka, M., Wong, K. C., More, A., *et al.* 2016, A Spectroscopically Confirmed Double Source Plane Lens System in the Hyper Suprime-Cam Subaru Strategic Program. *ApJL*, 826(2), L19.

Treu, T. 2010, Strong Lensing by Galaxies. *ARA&A*, 48, 87–125.

Tu, H., Gavazzi, R., Limousin, M., *et al.* 2009, The mass profile of early-type galaxies in overdense environments: the case of the double source-plane gravitational lens SL2SJ02176-0513. *A&A*, 501(2), 475–484.

Vegetti, S., Lagattuta, D. J., McKean, J. P., *et al.* 2012, Gravitational detection of a low-mass dark satellite galaxy at cosmological distance. *Nature*, 481(7381), 341–343.

Wang, H., Cañameras, R., Caminha, G. B., *et al.* 2022, Constraining the multi-scale dark-matter distribution in CASSOWARY 31 with strong gravitational lensing and stellar dynamics. *arXiv e-prints*, arXiv:2203.13759.

Zitrin, A., Rosati, P., Nonino, M., *et al.* 2012, CLASH: New Multiple Images Constraining the Inner Mass Profile of MACS J1206.2-0847. *ApJ*, 749(2), 97.

Strong Gravitational Lensing in the Era of Big Data
Proceedings IAU Symposium No. 381, 2024
H. Stacey, A. Sonnenfeld & C. Grillo, eds.
doi:10.1017/S1743921323003642

Large data set of lensed quasars: higher accuracy on H_0? The angular structures viewpoint

Lyne Van de Vyvere

Univerity of Liège STAR Institute, Quartier Agora - Allée du six Août, 19c B-4000 Liège, Belgium, email: lyne.vandevyvere@uliege.be

Abstract. Thanks to forthcoming large-scale surveys, a tremendous number of strong lenses will be discovered in the coming years. The gain in accuracy on H_0 from such a large population of lensed quasars is a key question for the future of time-delay cosmography. In such context, lensed systems will have to be modeled in an automated way, with models that are sufficiently generic to apply to every lens. I explore the biases that may arise from unaccounted-for azimuthal structures in mass models. The non-modeled twists in lensing galaxies are expected to bias the shear inference but not H_0. Disregarded ellipticity gradients, boxyness and discyness may impact the cosmological inference on a lens-by-lens basis. Nevertheless, the diversity of azimuthal mass profile in lenses balances the bias at a population level and the H_0 inference can thus benefits from such large surveys.

Keywords. gravitational lensing, galaxies: elliptical and lenticular, cD, galaxies: statistics, cosmological parameters, methods: numerical

1. Introduction

Time-delay strong gravitational lensing is a powerful tool to measure the Hubble constant, H_0. As a massive galaxy deforms space-time, light-rays coming from a background source are bent when passing by this massive object called the lens, and an observer sees several images of the source when the configuration allows it. Temporal variations of the source luminosity reach the observer at different times depending on the path the light followed. By measuring this time-delay and modeling the mass of the lens, one can infer cosmological parameters without relying on the distance-ladder, unlike most other methods.

Current H_0 inferences based on time-delay gravitational lensing combine the results from very accurate modeling of $\mathcal{O}(10)$ lensed quasars (Wong et al. 2020), out of the $\mathcal{O}(100)$ known lensed quasars. Forthcoming large-scale surveys, as Euclid, the Rubin Observatory, the Roman Space Telescope and the Chinese Space Station Telescope, among others, are expected to discover $\mathcal{O}(10^5)$ lenses. Such drastic increase of the available sample should enable a more accurate H_0 inference, but it will also force the lensing community to automatized the modeling, using models sufficiently generic to be applied to every lens.

Most lenses are massive elliptical galaxies. While in first approximation those galaxies can be considered as simple ellipsoids displaying the same position angle and projected ellipticity at all radii, elliptical galaxies are known to be more complex. Changes of position angle and ellipticity with galacto-centric radii are observed in isophotal profiles of elliptical galaxies (Kormendy et al. 2009). Moreover, deviations from ellipticity such as discyness and boxyness, that is multipoles of order 4, appear in 90% of local elliptical galaxies (Hao et al. 2006). These azimuthal structures, that is discyness, boxyness, twist, and ellipticity variations, have also been detected in the light profiles of elliptical galaxies at redshifts similar to lensing galaxies (Pasquali et al. 2006; Mitsuda et al. 2017; Keeton et al. 2000). Their origin can be various, ranging from projection effects of triaxial galaxies to formation and merging history of galaxies (Kormendy et al. 2009; Khochfar & Burkert 2005). As these azimuthal structures have been observed in the light of elliptical galaxies, one can advocate that such structures are also present in the mass of these galaxies, at least to some extend.

Currently, the presence of azimuthal structures in the light of lenses and the effect of such structures in the mass on the lensed images are at the limit of being detectable on image taken by the Hubble Space Telescope (HST). Few authors have been purposely accounting for azimuthal structures in the lens mass model of extended lensed images (e.g. Keeton et al. 2000; Powell et al. 2022). Nevertheless, some modeling software with pixelated lensing potential corrections or modeling strategies with multiple lens components will naturally allow for more azimuthal freedom. In the upcoming big data era, most models will not account for such freedom. The biases that may arise from unaccounted-for structures in mass models need to be explored, as well as their impact on time-delay cosmography.

2. Method

To explore the impact of unaccounted-for structures in the mass of lenses on the H_0 inference, the following experiment, which can be summarized in 4 steps, has been designed. First, mass profiles with azimuthal structures, that is, boxyness, discyness, twist, and ellipticity gradient, are emulated. Second, these mass profiles are used to lens a background quasar and its host galaxy, creating lensed images with HST-like data quality. Third, these images are modeled assuming that the mass profile is perfectly elliptical, i.e. displaying only 1 position angle and ellipticity. Fourth, the impact of such modeling on the cosmological inference is analysed, distinguishing the case-by-case inference from the population one.

To emulate azimuthal structures, different strategies have been used for the different types of structures. **Boxyness** or **discyness** can be emulated by adding an analytical multipolar component of order 4 (aka an octupolar momentum) on an existing elliptical profile. The distribution of the strength of octupolar deformations can be assessed using Hao et al. (2006) analyses of multipolar components in the light of elliptical galaxies, and restricting ourselves to pure boxyness or pure discyness. Mitsuda et al. (2017) found that for the most massive galaxies, discy and boxy galaxies can be found in a fifty-fifty proportion. With such information, a realistic population of lensing galaxies displaying boxyness or discyness can be created. **Position angle and ellipticity variations** with galacto-centric radii can be emulated using the method of Schramm (1994): the total mass of a galaxy can be approximated by a sum of finite and elliptical slices, each having a constant surface density and a given position angle and ellipticity. The total mass of the galaxy is a simple superposition of the slices, like a wedding cake is made of the superposition of tiers. Each slice can be analytically described, hence creating a smooth lensing potential, without artifacts (Van de Vyvere et al. 2020). To define the azimuthal structure of the profiles (aka the position angle and ellipticity variations), both hydrodynamical simulations of mass profile of massive elliptical galaxies (EAGLE, Schaye et al. 2015; Crain et al. 2015; McAlpine et al. 2016) and observations of light profiles of local elliptical galaxies (Kormendy et al. 2009) have been used.

The creation of mock lensing systems is performed with the `lenstronomy` software (Birrer et al. 2015; Birrer & Amara 2018; Birrer et al. 2021) in a standard way. No lens

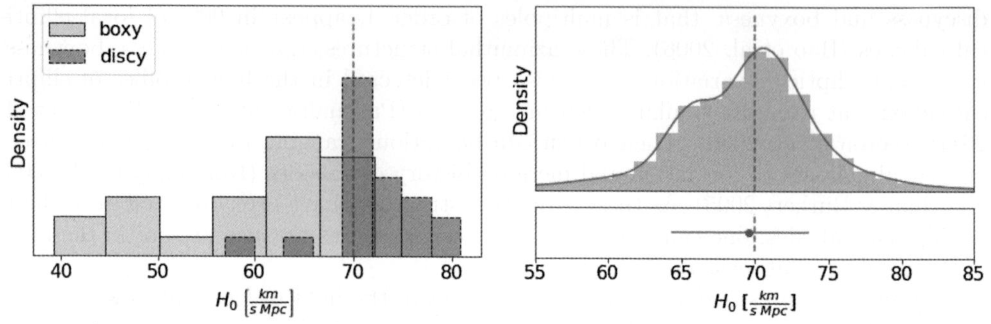

Figure 1. *Left:* Median H_0 inferences for random strength of multipolar deviations (within plausible range). *Right:* Population H_0 inference assuming equal proportion of boxy and discy galaxies, and strength of multipolar deviations based on observations of elliptical galaxies. The dashed blue vertical line indicates the fiducial H_0 used when emulating the lenses. Based on Van de Vyvere et al. (2022a)

light is created to focus on the effect of multipolar variations on the lensed images of the quasar and its host galaxy with HST-like data quality. Different Einstein radii and different radial mass profile prescriptions (i.e. isothermal power-law and composite mass model, made of Chameleon (Suyu et al. 2014; Dutton et al. 2011) and NFW (Navarro et al. 1996)) have been emulated. A shear component, emulating possible environmental effects, is also included in the mocks.

The mock images are modeled considering a power-law + shear model, without additional azimuthal structure. A cosmological inference is allowed thanks to the simulated time-delays.

A more detailed version of the methodology can be found in Van de Vyvere et al. (2022a,b).

3. Results

3.1. *Boxyness / discyness*

The multipoles in the lens mass affect mainly the arcs, aka the lensed image of the extended source. Such effect can be absorbed as a source feature in the model in which the mass can not display the boxy/discy azimuthal freedom. Nevertheless, with high signal-to-noise (i.e. best HST data available nowadays) and a high strength of the multipolar perturbation (strength typical of the utmost 10% of the elliptical galaxies population), the reconstructed source can look unphysical and the model would thus be flagged as unsuccessful.

For lenses analysed one by one, the H_0 inference can be biased, as can be seen in the left panel of Figure 1. Boxy lensing galaxies have a tendency to bias H_0 low when they are modeled as perfect ellipsoids, while the discy ones often bias H_0 high in the same context.

As boxyness and discyness effects on H_0 are opposite, the diversity of lenses allows an unbiased H_0 inference at a population level, as can be seen in the right panel of Figure 1. This result is true provided that the population is well represented by a 50-50 percentage of discy versus boxy galaxies, as measured by Mitsuda et al. (2017) for massive elliptical galaxies, and by a distribution of strength of multipolar deviations similar to the one measured by Hao et al. (2006). The upcoming time-delay cosmography in the big data era will benefit from studying selections effects in lenses (Sonnenfeld et al. 2023).

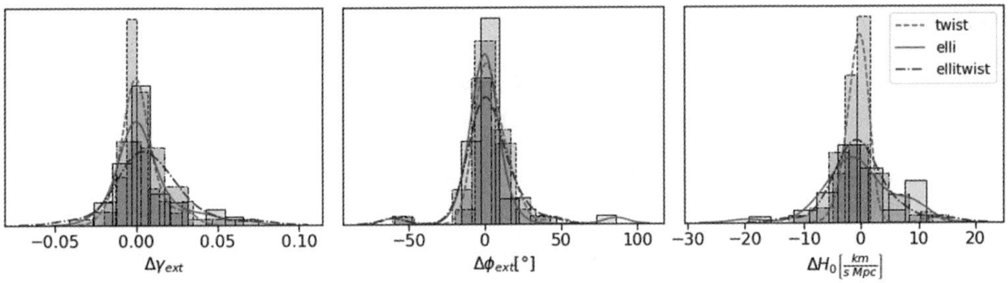

Figure 2. *Left:* Difference between input and modeled shear strength. *Middle:* Difference between input and modeled shear orientation. *Right:* Difference between input and modeled H_0. The variations in the mass of the mock lenses are based on both observed and hydro-simulated variations of position angle (dashed red and dash-dotted blue) and ellipticity (plain green and dash-dotted blue). Based on the results published in Van de Vyvere et al. (2022b)

3.2. *Twist / Ellipticity variations*

Mock images, created with twist and/or ellipticity variations in the mass, are easily modeled with perfect ellipsoids as lens mass models: the residuals are almost always compatible with pure noise and the modeled parameters are not unphysical. Nevertheless, several modeled parameters differ from the fiducial input values. The shear is one of them, as can be seen on the left and middle panels of Figure 2. This implies more generally that the modeled shear in real lenses may not reflect the environment when the lens mass model lacks azimuthal freedom, as also suggested by Etherington et al. (2023).

Unmodeled variations of position angles, aka twists, in the mass of lensing galaxies do not impact the H_0 inference, while the ellipticity ones do have an impact, as can be seen in the right panel of Figure 2. Even if the bias can reach $\sim$10 km s^{-1} Mpc^{-1} in the worst cases, the bias averages out at the population level. The latter being based on azimuthal profiles of both observations of nearby galaxies (Kormendy et al. 2009) and hydro-simulations of galaxies at typical lens redshifts (Schaye et al. 2015; Crain et al. 2015; McAlpine et al. 2016).

4. Conclusion and perspectives

As lenses will be uniformly modeled in the upcoming big data era, one may wonder if existing azimuthal structures in lenses can bias the cosmological inference if the lenses are modeled as simple ellipsoids. We explored the role of boxyness, discyness, twists, and ellipticity variations. Based on motivated samples of the different azimuthal structures, mock lenses displaying those structures were created and the resulting lensed images were modeled assuming that the lens mass is a perfect ellipsoid. In a nutshell, modeling lens images with a simple ellipsoid as the main lens can bias the H_0 inference on a case-by-case basis but the bias averages out at the population level considered. Nevertheless, some parameters as the source structure or the shear inference can be highly impacted and their general interpretation must be done carefully. The complete methodology and results can be found in Van de Vyvere et al. (2022a,b).

The selection effects in the upcoming new sample of lenses will need to be explored to ensure robust time-delay cosmography. The constraints that resolved kinematics can bring on assessing the structure of lenses has not yet been assessed, but is promising (Cappellari 2016) and should be investigated in the future.

On case-by-case basis, modeling the azimuthal structures should be feasible, assuming that high quality data are available. Keeton et al. (2000), Galan et al. (2022), and Powell et al. (2022) already made the first steps in that direction and a more general methodology combining the modeling of several azimuthal structures is still pending.

References

Birrer, S. & Amara, A. 2018, Physics of the Dark Universe, 22, 189

Birrer, S., Amara, A., & Refregier, A. 2015, *ApJ*, 813, 102

Birrer, S., Shajib, A. J., Gilman, D., et al. 2021, *JOSS*, 6, 3283

Cappellari, M. 2016, *ARA&A*, 54, 597

Crain, R. A., Schaye, J., Bower, R. G., et al. 2015, *MNRAS*, 450, 1937

Dutton, A. A., Brewer, B. J., Marshall, P. J., et al. 2011, *MNRAS*, 417, 1621

Etherington, A., Nightingale, J. W., Massey, R., et al. 2023, arXiv e-prints, arXiv:2301.05244

Galan, A., Vernardos, G., Peel, A., Courbin, F., & Starck, J.-L. 2022, *A&A*, 668, A155

Hao, C. N., Mao, S., Deng, Z. G., Xia, X. Y., & Wu, H. 2006, *MNRAS*, 370, 1339

Keeton, C. R., Falco, E. E., Impey, C. D., et al. 2000, *ApJ*, 542, 74

Khochfar, S. & Burkert, A. 2005, *MNRAS*, 359, 1379

Kormendy, J., Fisher, D. B., Cornell, M. E., & Bender, R. 2009, *ApJS*, 182, 216

McAlpine, S., Helly, J. C., Schaller, M., et al. 2016, Astronomy and Computing, 15, 72

Mitsuda, K., Doi, M., Morokuma, T., et al. 2017, *ApJ*, 834, 109

Navarro, J. F., Frenk, C. S., & White, S. D. M. 1996, *ApJ*, 462, 563

Pasquali, A., Ferreras, I., Panagia, N., et al. 2006, *ApJ*, 636, 115

Powell, D. M., Vegetti, S., McKean, J. P., et al. 2022, *MNRAS*, 516, 1808

Schaye, J., Crain, R. A., Bower, R. G., et al. 2015, *MNRAS*, 446, 521

Schramm, T. 1994, *A&A*, 284, 44

Sonnenfeld, A., Li, S.-S., Despali, G., Shajib, A., & Taylor, E. 2023, arXiv e-prints, arXiv:2301.13230

Suyu, S. H., Treu, T., Hilbert, S., et al. 2014, *ApJL*, 788, L35

Van de Vyvere, L., Gomer, M. R., Sluse, D., et al. 2022a, *A&A*, 659, A127

Van de Vyvere, L., Sluse, D., Gomer, M. R., & Mukherjee, S. 2022b, *A&A*, 663, A179

Van de Vyvere, L., Sluse, D., Mukherjee, S., Xu, D., & Birrer, S. 2020, *A&A*, 644, A108

Wong, K. C., Suyu, S. H., Chen, G. C. F., et al. 2020, *MNRAS*, 498, 1420

Strong Gravitational Lensing in the Era of Big Data
Proceedings IAU Symposium No. 381, 2024
H. Stacey, A. Sonnenfeld & C. Grillo, eds.
doi:10.1017/S174392132300426X

Accounting for population-level systematic effects using a hierarchical strategy

Matthew R. Gomer

University of Liège. email: matthewrgomer@gmail.com

Abstract. One of the largest sources of systematics in time-delay cosmography arises from Mass Sheet Transformation (MST). The degeneracy associated with this transformation is often broken by an assumed profile shape, such as a power-law. A hierarchical strategy has been developed which constrains the global profile shape on a population level, constrained collectively by the kinematics measurements of the lenses. This framework allows one to include non-time-delay lenses to provide constraints to the global profile, improving the H_0 constraints. This work tests the hierarchical framework using analytical profiles, and additionally tests the capacity to combine two populations which come from the same profiles but probe different radii due to a change in source redshift. We find that the hierarchical framework is able to compensate for this effect, and the addition of non-time-delay lenses improves the H_0 constraint, even though these lenses have different Einstein radii than their time-delay counterparts.

Keywords. Gravitational lensing: strong, cosmological parameters, Galaxies: statistics

1. Introduction

This article has been adapted from "TDCOSMO VIII: A key test of systematics in the hierarchical method of time-delay cosmography" (Gomer *et al.* 2022). For more details, we refer the reader to the original work.

The goal of time-delay cosmography is to use gravitational lensing to measure cosmological distances, and hence infer cosmological parameters such as H_0. Typically, the lens is an early-type galaxy and the source is a quasar. For a given system, the time delay between multiple images of the source is measured and compared against that predicted by the lens model, giving a measure of the so-called time-delay distance $D_{\Delta t} \propto H_0^{-1}$. Perhaps the largest source of systematics in such an analysis is due to the Mass Sheet Transformation (MST, Falco *et al.* 1985; Schneider & Sluse 2013):

$$\kappa(\theta) \rightarrow \lambda\kappa(\theta) + (1 - \lambda). \tag{1}$$

For any value of λ, the transformed convergence will reproduce the lensed imaging information and as such λ cannot be constrained by lensing alone. The time delays, and consequently the determination of H_0, are affected by a factor of λ, meaning an accurate determination of H_0 requires a constraint on λ coming from information external to the lensing data. Two sources of such information are possible: an assumed model profile shape (e.g. assumed λ) based on the known structure of early-type galaxies, and/or a measurement of the dynamics of the lens galaxy, which provides a direct measure of mass within some aperture, constraining the physically allowed value of λ. In this work, we restrict ourselves to the internal MST related to profile shape, specified with the notation that $\lambda = \lambda_{\rm int}$.

To date, the most precise measures of H_0 using time-delay cosmography come from the TDCOSMO collaboration (Millon *et al.* 2020). The process consists of jointly modeling lens imaging data and time delays, along with using kinematic measurements to account for the MST. Using a parametric form for the mass profile in the form of a power-law model, H_0 can be inferred to $\sim 2.5\%$ precision by combining the results from 6 systems (Wong *et al.* 2020). However, this result does not completely account for the full flexibility of the MST, as much of the constraint on $\lambda_{\rm int}$ is informed by the power-law assumption. In order to include this systematic effect in the total error budget, it must be accounted for in the modeling procedure.

1.1. *A hierarchical strategy*

Birrer *et al.* (2022) introduced a technique to supplement the parametric method by explicitly folding the effect of the MST into a hierarchical analysis of the lens systems. The approach applies an MST to the lens modeling results on a population level, serving to capture systemic departure from the initial parametric model.

The hierarchical strategy assumes that as a population, the lens galaxies have some global profile shape which is within an MST of a power law. As such, the $\lambda_{\rm int}$ experienced by a given power-law lensing fit is a function of where on this global profile the Einstein radius ($\theta_{\rm Ein}$), lies. Meanwhile, the kinematic measurements of a given system measure the absolute mass within some aperture size, which has the power to constrain $\lambda_{\rm int}$. Therefore, each lens brings a probe of this global mass distribution, with kinematics information at one radius and lensing information at another. The hierarchical framework combines many systems together, allowing for variation within the population, to measure a population-scale $\lambda_{\rm int}$.

More explicitly, this framework ascribes a distribution of MSTs to the galaxy population, where the distribution is described with Gaussian standard deviation $\sigma(\lambda_{\rm int})$. The transformation is allowed to vary as a function of where in the profile the Einstein radius lies, relative to the effective radius of the galaxy light distribution, $\theta_{\rm eff}$. As the exact function of this transformation is unknown, it is approximated as linear with slope α_λ and intercept $\lambda_{\rm int,0}$:

$$\lambda_{\rm int}(\theta_{\rm eff}/\theta_{\rm Ein}) = \lambda_{\rm int,0} + \alpha_\lambda \left(\frac{\theta_{\rm eff}}{\theta_{\rm Ein}} - 1\right), \tag{2}$$

where $\lambda_{\rm int}$ is the internal MST experienced by a given lens. In addition, two more population-scale parameters are included describing the anisotropy of the kinematics of lens galaxies: $\langle a_{\rm ani}\rangle$ and $\sigma(a_{\rm ani})$ describe the mean and Gaussian width of the anisotropy radius of the lens systems, scaled as $a = r_{\rm ani}/r_{\rm eff}$. Applying this framework to 7 TDCOSMO systems, (Birrer *et al.* 2022) found $H_0 = 74.5^{+5.6}_{-6.1}$ km s^{-1} Mpc^{-1}, where the uncertainty now accounts for the MST.

The beauty of this strategy is that its assumptions on the profile shape are data-driven, and as such it can benefit from including additional data. Lens models and kinematics from other systems can be incorporated into the framework to provide better constraints on the profile shape, even if they themselves lack time delays and cannot be used for H_0 determination. The inherent assumption in this practice is that all lens galaxies included in the inference come from the same global population. Birrer *et al.* (2022) combined 33 SLACS lenses with the 7 TDCOSMO results, finding $H_0 = 67.4^{+4.1}_{-3.2}$ km s^{-1} Mpc^{-1}.

The results are consistent with one another, although the mean value has shifted. The main concern with this analysis is that if the SLACS systems are systematically different than the TDCOSMO systems, combining them could bias the resulting value of H_0. The most obvious difference between these systems is that they have different redshifts and therefore probe different parts of the profile. In theory, the linear scaling with radius should account for this change.

This sets the stage for a targeted test of systematics in the hierarchical strategy. In this work, we will use mock lenses within the same population for which different parts of the profile are probed by lensing due only to a change in redshift. Modeling and combining these systems hierarchically provides a test to see if the hierarchical framework can account for this effect.

2. Experiment setup

We test the hierarchical framework using a population of mock lenses with analytical mass distributions. This mock population resembles the SLACS lenses (Shajib *et al.* 2021) when the source is placed at a redshift of $z_s = 0.6$, and resemble the TDCOSMO lenses when the source is placed at a redshift of $z_s = 2.0$. These choices result in critical densities for each set which is consistent with its respective population.

Each lens is described using two components: a Chameleon profile emulates a Sérsic profile to represent the stellar mass distribution, while an NFW component represents the dark matter distribution. To determine the parameters for the Chameleon and NFW profiles, we sample a large range of values and select the resulting profiles which match the observed populations by several metrics, namely the Einstein radius $\theta_{\rm Ein}$, when placed at both $z_s = 0.6$ and $z_s = 2.0$, the effective radius of the light distribution $\theta_{\rm eff}$, and the axis ratio q. Selecting systems which match these criteria give us a set of systems for which we can create a "SLACS-like" system when placing the source at $z_s = 0.6$, and a "TDCOSMO-like" counterpart for which the only change is that the source is now at $z_s = 2.0$.

We randomly select 20 pairs of "SLACS-like" and "TDCOSMO-like" mock profiles to test the hierarchical framework. First, we create mock images from these profiles using `lenstronomy` (Birrer *et al.* 2018, 2021a), with resolution and noise settings matching the F160W filter on the HST WFC3 camera. The mock source is a circular Sérsic source, and the mock lens light is not included so as to emulate a perfect light subtraction. The sources for the TDCOSMO-like systems include a point source which emulates a quasar. The TDCOSMO-like systems are also given mock time-delay measurements. Each mock is fit using Power-Law Elliptical Mass Distribution (PEMD) + external shear model using `lenstronomy`. All mock images are successfully fit to the level of the noise, with a reduced $\chi^2 < 1.06$ in all cases.

Mock kinematics are calculated using spherical Jeans kinematics, in which the velocity dispersion is calculated within an aperture under the assumption of a relaxed system with spherical symmetry. Mock kinematic measurements are calculated for three different apertures within 2" and given an uncertainty of 10%, emulating spatially-resolved kinematics observations. For the truth anisotropy, we set $a_{\rm ani} = 1$.

The systems are then combined hierarchically, using the `hierArc`† package. Input into the framework are the posteriors from the lens modeling as well as the kinematics measurements. The framework then recovers the population-scale $\lambda_{\rm int}$ according to Eq. 2. Aperture kinematics scale with $\sqrt{\lambda_{\rm int}}$, thereby informing the nature of the transformation from the power-law models to the global profile shape. Ultimately, posteriors are recovered which explicitly include the effect of $\lambda_{\rm int}$ in the error budget of H_0.

3. Results and Discussion

Before combining the systems hierarchically, the individual power-law fits recover values of H_0 which are biased by approximately 15%, due to the MST. For this work, we combined hierarchically the TDCOSMO-like lenses alone, the SLACS-like lenses alone,

† `https://github.com/sibirrer/hierArc`

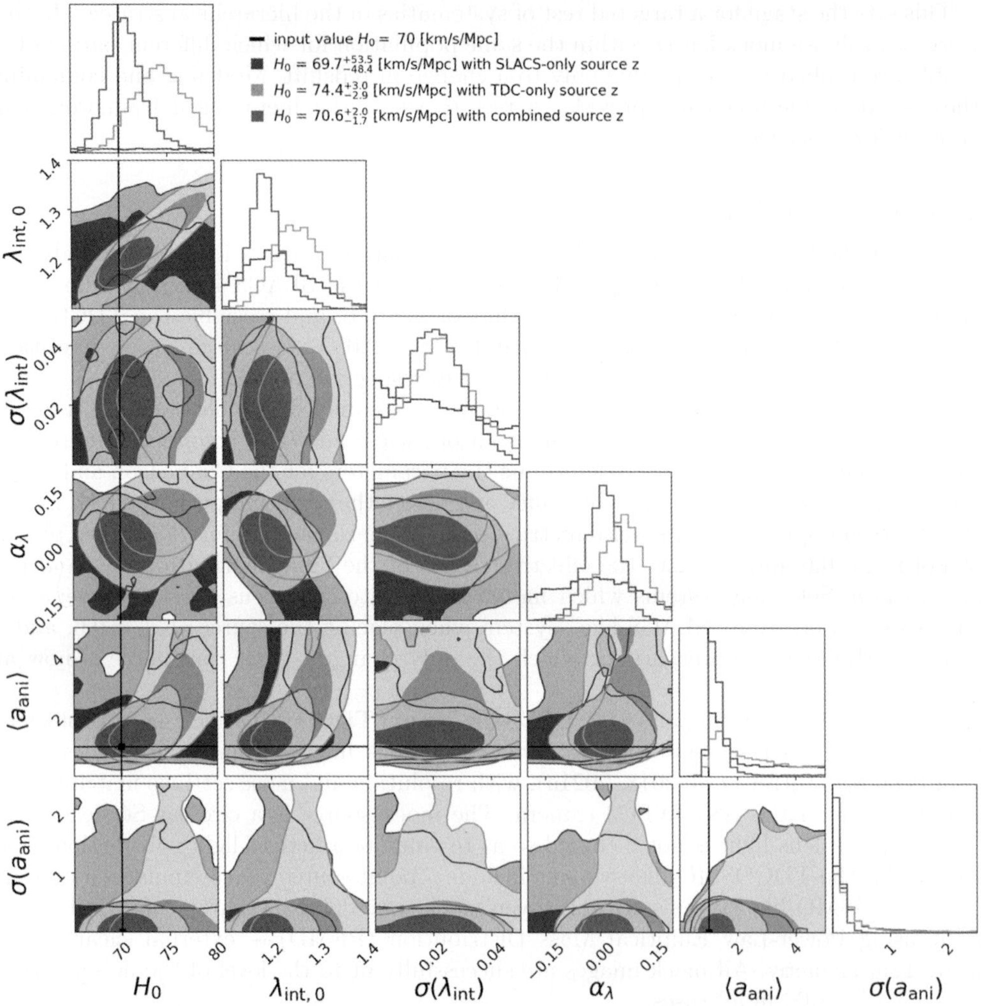

Figure 1. Corner plot for the results of the hierarchical test. Orange: TDCOSMO-like lens population; Purple: SLACS-like population; Blue: Both populations combined. Credit: Gomer *et al.* (2022).

and the combination of the TDCOSMO-like and SLACS-like systems together. These results are shown in Fig. 1.

The TDCOSMO-only result recovers an H_0 of $74.4^{+3.0}_{-2.9}$ km s^{-1} Mpc^{-1}. This posterior is within 1.5σ of the fiducial value of 70 km s^{-1} Mpc^{-1}, with the median now within 6% of the fiducial value. This improvement in accuracy compared to the nonhierarchical case comes from the explicit treatment of $\lambda_{\rm int}$, which is recovered to have a value different than 1.

The result from the SLACS only lens puts constraints on parameters such as $\lambda_{\rm int}$ and $\langle a_{\rm ani} \rangle$, but since the SLACS-like systems lack time delays they cannot constrain H_0 alone. However, the combination of the two populations hierarchically provides the strongest result, recovering H_0 of $70.6^{+2.0}_{-1.7}$ km s^{-1} Mpc^{-1}, closely matching the fiducial value. The additional constraints on the parameters other than H_0 informs the population-scale MST mapping, allowing an accurate recovery of H_0.

Additional tests with a larger sample size found that this precision continues to increase with more systems. This means that the intrinsic scatter in $\lambda_{\rm int}$ is captured by the hierarchical framework and does not pose a systematic floor to the achievable H_0 precision.

3.1. *A look ahead for hierarchical strategies*

The hierarchical framework has been shown to adequately account for the most well-studied lensing degeneracy: the MST. However, there are other types of degeneracies that can be a nuisance in lens modeling. Van de Vyvere *et al.* (2022a,b) explored the effect of azimuthal structure on time-delay cosmography, finding that structures such as disciness, boxiness, and ellipticity gradients can bias the recovered value of H_0 for individual systems, although as a population this effect seems to average out.

We have seen that a hierarchical strategy can allow one to probe the structure beneath the statistical scatter by considering a population, and so naturally one wonders if this could be an applicable strategy to approach this problem. The main challenge is to express an analytical description of these complex degeneracies, which would map a given azimuthal perturbation to the effect it has on H_0 for an individual system. Such a description could then be applied hierarchically to a population, recovering the expected azimuthal contribution on a population scale. This is an ongoing field for exploration for lensing theory (see also Sonnenfeld 2017; Wagner 2019; Kochanek 2021; Birrer 2021b; Gomer *et al.* 2023). Despite its theoretical challenges, such a method may be the way forward for time-delay cosmography in the future, particularly in the big data era where there are too many lenses to model in great detail.

4. Conclusion

Perhaps the largest source of systematic errors in time-delay cosmography comes from an uncertainty on the profile shape, through the Mass Sheet Transformation (MST). A hierarchical framework has been implemented which accounts for the MST on a population level, constrained by the kinematics of many lens systems. In this work, we test this hierarchical framework using analytical profiles, and find that it reduces the bias associated with the MST from $\sim 15\%$ to $\sim 6\%$.

The strength of the hierarchical strategy is that additional systems can be added to inform the MST at the population level, even if they do not directly constrain H_0 alone. Birrer *et al.* (2022) combined SLACS and TDCOSMO systems together using this framework, although one possible concern is that if these systems come from different populations, this could return a biased result. We perform a targeted systematics test by creating mock SLACS-like lenses which have the same profiles as their TDCOSMO-like counterparts, except that a change in source redshift changes the Einstein radius. We show that the hierarchical framework is able to accommodate this systematic change, and that the inclusion of these SLACS-like systems results in a more accurate H_0. We note however that any evolution in the global properties of the profile has not been tested by this work: to capture such trends may require that a parameterization of such trends be included in an updated hierarchical implementation. Nevertheless, this result strengthens the argument for a hierarchical strategy, which we speculate will become a predominant method for time-delay cosmography in the era of big data.

References

Birrer, S. and Amara, A. 2018, *Physics of the Dark Universe*, 22, 189
Birrer, S., Shajib, A., Gilman, D. *et al.* 2021, *The Journal of Open Source Software*, 6, 62
Birrer, S. 2021, *ApJ*, 919, 38

Birrer, S., Shajib, A. J., Galan, A. *et al.* 2022, *A&A*, 643, A165
Falco, E. E., Gorenstein, M. V. and Shapiro, I. I. 1985, *ApJ*, 289, L1-L4
Gomer, M. R., Sluse, D., Van de Vyvere, L., *et al.* 2022, *A&A*, 667, A86
Gomer, M. R., Sluse, D., Van de Vyvere, L., *et al.* 2023, *(in prep.)*
Kochanek, C. S. 2021, *MNRAS*, 501, 5021
Millon, M., Galan, A., Courbin, F., *et al.* 2020, *A&A*, 639, A101
Schneider, P. and Sluse, D. 2013, *A&A*, 559, A37
Shajib, A. J., Treu, T., Birrer, S., *et al.* 2021, *MNRAS*, 503, 2380
Sonnenfeld, A. 2017, *MNRAS*, 474, 4648
Van de Vyvere, L., Sluse, D., Mukherjee, S., *et al.* 2022, *A&A*, 659, A127
Van de Vyvere, L., Sluse, D., Gomer, M. R., *et al.* 2022, *A&A*, 663, A179
Wagner, J. 2019, *MNRAS*, 487, 4492
Wong, K. C., Suyu, S. H., Chen, G. -F., *et al.* 2020, *MNRAS*, 498, 1420

Strong Gravitational Lensing in the Era of Big Data
Proceedings IAU Symposium No. 381, 2024
H. Stacey, A. Sonnenfeld & C. Grillo, eds.
doi:10.1017/S1743921323004362

Time Delay Cosmograpy of SDSSJ1433 with the 2.1m Wendelstein Telescope

G. Queirolo

Universitäts-Sternwarte, Fakultät für Physik, Ludwig-Maximilians Universität München, Scheinerstr. 1, 81679, München, Germany
Max Planck Institute for Extraterrestrial Physics, Giessenbachstrasse 1, 85748 Garching, Germany. email: queirolo@usm.lmu.de

Abstract. Time delay cosmography is based on the study of multiply gravitationally lensed images of a variable source. Their time delay differences are linearly dependent on the Fermat potential differences at the images' positions and the Hubble parameter, whose exact value is to this date strongly debated in the framework of the *Hubble tension.* In this paper we present the study of SDSSJ1433, a quadruply lensed QSO, the time delays of which have been obtained after a 3-year observational campaign from the 2.1m Wendelstein telescope in the optical g' filter, and the corresponding mass model was constrained from multi-band archival *HST* observations. The resulting H_0 value is $77.4\frac{km}{s\,Mpc}$ with a precision of $\sim 6\%$.

Keywords. Gravitational lensing, cosmology, galaxies, Hubble constant

1. Introduction

Strong gravitational lens systems offer the opportunity to measure gravitationally induced time delays in intrinsically varying sources. By combining these delays with mass-modelled Fermat potentials, the Hubble parameter can be constrained (Refsdal (1964)). The precision of individual systems depends on the modelling choices of lens features, source variability, photometric precision and cadence of the monitoring. Thus it can vary, in the best cases, between 2.5% and 10%. A combination of 7 such systems yielded a formal Hubble constant error of 2.4% and a value of $73.3\frac{km}{s\,Mpc}$ (Wong, et al. (2020)). Hubble parameter estimates can be made more accurate if profiles of elliptical galaxies (fulfilling the selection function of strong lensing systems) are known more precisely and if the numbers of such systems with accurate time delay measurements can be increased.
Therefore this project, following the path of the **H0LiCOW** collaboration, was envisioned to study the system SDSSJ1433+6007, a quadruply lensed QSO, and further test the methodology by following an independent implementation.

2. Lensing Model

To constrain the differences of Fermat Potential between the images' positions I and J, $\Delta\phi_{IJ}$, the python library **Lenstronomy** (Birrer and Amara (2018)) was applied to *HST* archival observations in multiple filters; the modelling was done independently for each filter, given the same prior, and the posterior on $\Delta\phi_{IJ}$ were combined by multiplying their likelihoods (see Figure 1 and Table 1).

Table 1. Results for the difference of Fermat potential $\Delta\phi$ and the time delay Δt at the images' position (mean and 1-σ).

	$\Delta\phi$	Δt
AB	$0.43^{0.02}_{0.04}$ arcsec^2	22.3±2.2 d
AC	$0.63^{0.03}_{0.04}$ arcsec^2	33.7±2.5 d
AD	$1.48^{0.06}_{0.1}$ arcsec^2	81.1±3.5 d

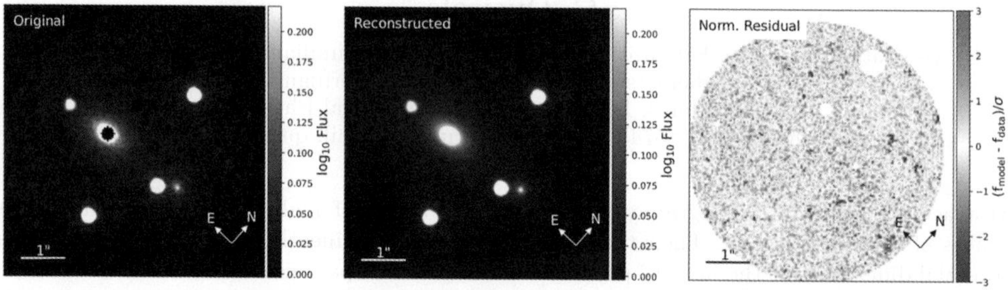

Figure 1. *HST* observation of SDSSJ1433 in the optical filter F814W, with the corresponding lens model and its residual.

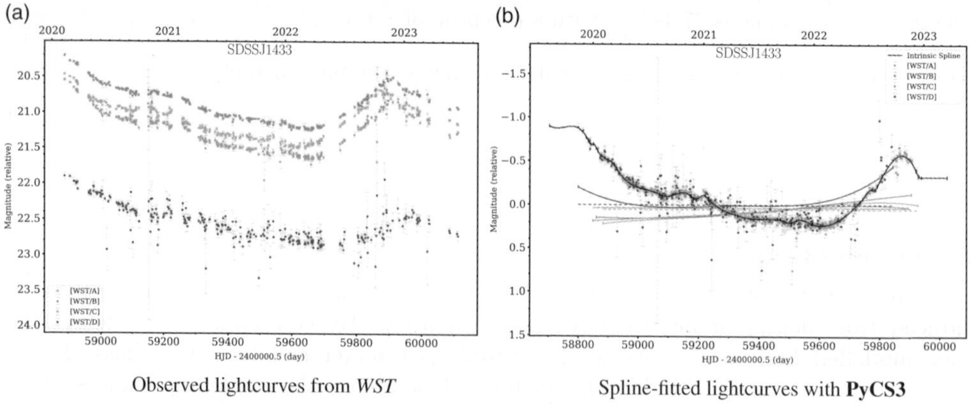

Figure 2. Time delay analysis of the lightcurves observed from *WST* with spline fitting.

3. Time Delay Measurement

The system was observed from the Wide Field Imager of the 2.1m Wendelstein Telescope (*WST*), surveying the QSO for over 3 years in the *g'* band. The final dataset is comprised of over 300 nights with an average cadence of 4 days and a median PSF of 1.07". Using our data reduction (Kluge, et al. (2020)) and difference imaging pipelines (Gössl and Riffeser (2002)) the lightcurves were obtained and then analysed with **PyCS3** (Millon, et al. (2020)) utilising the spline fitting method (see Figure 2).

4. H_0 Estimate

$P(H_0 \mid \Delta t_{\rm IJ}, \Delta\phi_{\rm IJ})$, the posterior on H_0 given the time delay posterior $P(\Delta t_{\rm IJ})$ and the Fermat potential difference posterior $P(\Delta\phi_{\rm IJ})$, is computed following Bayes' theorem. This yields the following equation:

$$P(H_0 \mid \Delta t_{\rm IJ}, \Delta\phi_{\rm IJ}) = \int d\Delta\phi_{\rm IJ} P(\Delta\phi_{\rm IJ}) \frac{P_{\Delta t\,\rm map}(\Delta\phi_{\rm IJ} \mid H_0) Prior(H_0)}{\int dH_0' P_{\Delta t\,\rm map}(\Delta\phi_{\rm IJ} \mid H_0')}$$

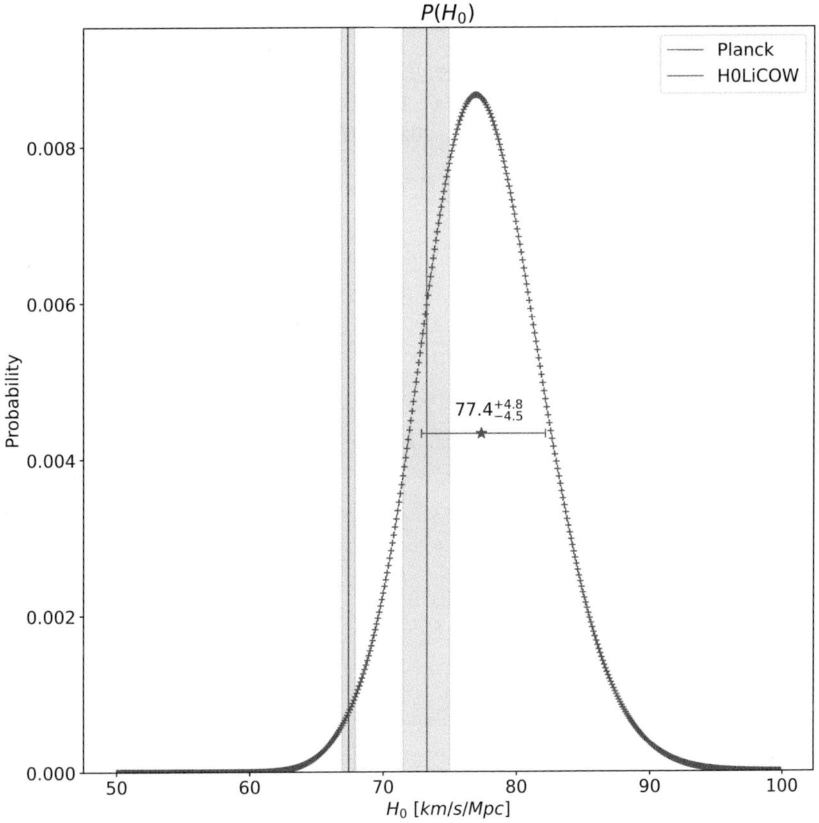

Figure 3. H_0 posterior compared with two characteristic results of the Hubble tension: Planck (Planck Collaboration (2020)) and H0LiCOW (Wong, et al. (2020)).

where $P_{\Delta \mathrm{t\,map}}(\Delta\phi_{\mathrm{IJ}} \mid H_0)$ is $P(\Delta t_{\mathrm{IJ}})$ which has been mapped into the $\Delta\phi_{\mathrm{IJ}}$ parametric space by assuming a value of H_0. This is then normalised over all possible values of H_0 and multiplied by $P(\Delta\phi_{\mathrm{IJ}})$, the posterior on $\Delta\phi_{\mathrm{IJ}}$ obtained from the lens modelling, and integrated over all $\Delta\phi_{\mathrm{IJ}}$. Finally a flat prior $Prior(H_0)$ is considered. Following this approach, the preliminary constraint on the Hubble constant yields $77.4\frac{km}{s\,Mpc}$ with a precision of $\sim 6\%$ (Queirolo, et al. (in prep.)) (see Figure 3). This computation assumes a flat ΛCDM and fixed, error free value of $\Omega_m = 0.30966$, in agreement with the cosmological result of the Planck Collaboration (2020). Possible biases, which are yet to be investigated, may arise from an unaccounted mass sheet degeneracy (Schneider and Sluse (2013)) or equivalently the assumed mass profile slope (Birrer, et al. (2020)).

References

Birrer, S., Amara, A. 2018, Physics of the Dark Universe, 22, 189

Birrer, S., Shajib, A.J., Galan, A., Millon, M., Treu, T., Agnello, A., Auger, M., Chen, G.C.-F., Christensen, L., Collett, T. 2020, *A&A* 643, A165

Gössl, C. A., Riffeser, A. 2002, *A&A* 381, 3, 1095

Kluge, M., Neureiter, B., Riffeser, A., Bender, R., Goessl, C., Hopp, U., Schmidt, M., Ries, C., Brosch, N. 2020, *ApJS* 247, 2, 43

Millon, M., Tewes, M., Bonvin, V., Lengen, B., Courbin, F. 2020, *JOSS* 5, 53, 2654

Planck Collaboration 2020, *A&A* 641, A12

Queirolo, G., Seitz, S., Riffeser, A., Kluge, M. *in prep.*
Refsdal, S. 1964, *MNRAS* 128, 4, 307
Schneider, P., Sluse, D. 2013, *A&A* 559, A37
Wong, K., Suyu, S.H., Chen, G. C.F., Rusu, C. E., Millon, M., Sluse, D., Bonvin, V., Fassnacht, C. D., Taubenberger, S., Auger, M. W. 2020, *MNRAS* 498, 1, 1420

Strong Gravitational Lensing in the Era of Big Data
Proceedings IAU Symposium No. 381, 2024
H. Stacey, A. Sonnenfeld & C. Grillo, eds.
doi:10.1017/S1743921323004246

Impact of Galaxy Dynamics on Modified Gravity Constraints from Strong Lensing Systems

Grasiele Romanzini-Bezerra[1], **Martin Makler**[2,1] **and Cristina Furlanetto**[3]

[1]Centro Brasileiro de Pesquisas Físicas, Rio de Janeiro, Brazil.
email: bezerragrasiele@gmail.com

[2]ICIFI, Universidad Nacional de San Martín, Provincia de Buenos Aires, Argentina

[3]Universidade Federal do Rio Grande Sul, Porto Alegre, Brazil

Abstract. Galaxy-galaxy strong lensing systems have been used in literature to test General Relativity by constraining the post-Newtonian parameter (γ_{PPN}). Nevertheless, these methods are prone to systematic errors, some of which arise from difficulties in modelling the dynamics of the lens galaxy. In this study, we address the systematic error related to the assumption of a constant anisotropy between the radial and tangential components of the velocity dispersion of stars within the lens galaxy, characterised by the parameter β. We considered two radial models for the anisotropy parameter, the Osipkov-Merritt and Mamon & Lokas, as well three Gaussian priors for constant β. Our analysis showed that the choice of β has a strong impact on the value of γ_{PPN}, with radial models leading to lower values of this parameter.

Keywords. Strong lensing, galaxies

A test of General Relativity (GR) can be carried out using galaxy-galaxy strong gravitational lensing (SL) systems. In the weak field approximation, it is possible to characterise the space-time metric with two potentials: a Newtonian, Φ, and a curvature one, Ψ. The ratio of these two potentials is the post-Newtonian parameter, γ_{PPN}. In GR the value for this post-Newtonian parameter is one.

It is possible to measure the post-Newtonian parameter at galactic scales by using SL systems in which the lens object is an early-type galaxy (ETG) (e.g., Cao *et al.* 2017). As gravitational lensing is sensitive to the sum of the potentials $(\Phi + \Psi)$, whereas the stellar kinematics exclusively relates to the Newtonian potential (Φ), combining these observables enables effective constraints on the parameter γ_{PPN}.

Despite the recent advancements in the measurement of γ_{PPN} through this method (Cao *et al.* 2017; Collett *et al.* 2018), there are still systematic uncertainties to be addressed. A specially concerning one is the systematic error caused by the incomplete knowledge on the dynamics of the lens galaxy. In particular, the anisotropy in the velocity dispersion of its stars, β, cannot be directly determined from observational data and is not accurately predicted from simulations. Addressing and understanding this systematic error is crucial for accurate measurements of γ_{PPN}.

In this work, we employ a methodology similar to that of Cao *et al.* (2017), where the velocity dispersion of the stars in the lens galaxy is determined by assuming power-law models for both the mass density and brightness profiles. The derived velocity dispersion is utilised to calculate $\overline{\sigma_*}(\theta_E)$, representing the observed line-of-sight velocity dispersion

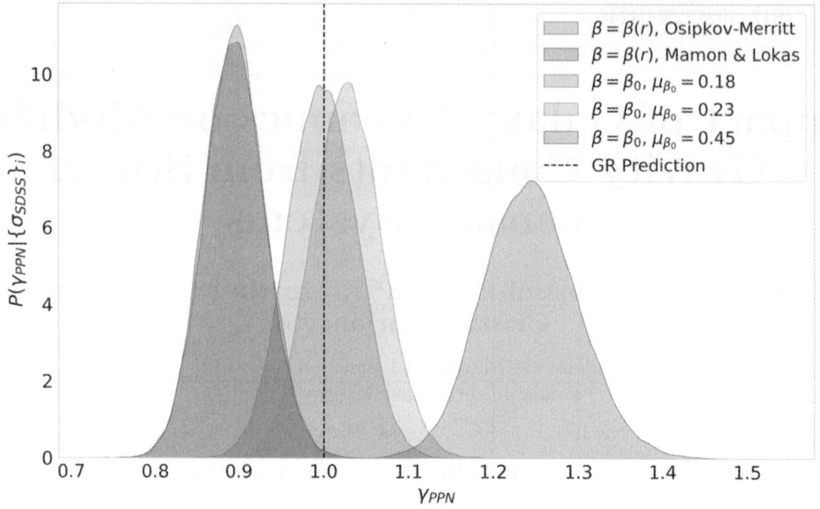

Figure 1. The posterior PDF of γ_{PPN} for the five models considered in this work: the two $\beta(r)$ models of OM and ML and the three Gaussian priors on β.

within the spectroscopic aperture. This computation takes into account the lens brightness and corrects for the point spread function. Furthermore, for this study we consider different priors on β and investigate their impact on the constraints on γ_{PPN}. First we derived the posterior probability density function (PDF) for γ_{PPN} for three Gaussian priors with the following parameters from the literature (written as: $\beta_0 = \mu_\beta \pm \sigma_\beta$): $\beta_0 = 0.18 \pm 0.13$, as in Cao *et al.* (2017); $\beta_0 = 0.45 \pm 0.25$, from Koopmans *et al.* (2009); and $\beta_0 = 0.23 \pm 0.19$, from Wang *et al.* (2019).

In addition to a constant velocity anisotropy, we consider two radial profiles for β: the Osipkov-Merritt (OM) model (Osipkov 1979; Merritt 1985) and the Mamon & Lokas (ML) one (Mamon *et al.* 2005). The OM is given by

$$\beta_{\rm OM}(r) = \frac{r^2}{r^2 + r_a^2}, \tag{1}$$

where $r_a = a\, r_{200}$, with r_{200} being the radius where the mean density of the galaxy is equal to 200 times the critical density of the universe. And the ML is given by

$$\beta_{\rm ML}(r) = \frac{1}{2}\frac{r}{r + r_a}. \tag{2}$$

We employed the aforementioned anisotropy profiles in the computation of $\overline{\sigma_*}(\theta_E)$. This quantity can be directly compared to the actual velocity dispersion σ_v obtained from the spectra of the lens galaxies of our sample. In addition to the Einstein radius, θ_E, measured from the SL modelling, $\overline{\sigma_*}$ depends on the density and brightness profiles, the cosmological distances involved, $\beta(r)$ and γ_{PPN}.

From the data set comprising 80 strong lensing systems used in Cao *et al.* (2017), containing measurements of θ_E and σ_v, we optimised the likelihood to obtain a PDF for γ_{PPN} for the different anisotropy models described above. The results are shown in Figure 1. The two superimposed curves on the left correspond to the OM and ML models for $\beta(r)$. The green curve following those reproduces the result of Cao *et al.* (2017), which used $\beta_0 = 0.18 \pm 0.13$. The two curves on the right (orange and blue), correspond, respectively, to the priors $\beta_0 = 0.23 \pm 0.19$ and $\beta_0 = 0.45 \pm 0.25$.

In summary, there is a clear correlation between β_0 adopted and γ_{PPN}. Furthermore, the models featuring radially dependent anisotropy profiles (OM and ML) yield lower values for the post-Newtonian parameter values than the constant β models considered. The pivotal inference drawn from this analysis underscores the pronounced influence of assumptions surrounding β on the resultant γ_{PPN} outcomes, thereby evoking considerations about potential biases in prior selections for β in preceding investigations. Despite the significance of contrasting lens stellar kinematics with the lensing phenomenon as a crucial experiment for GR and modified gravity theories at galactic scales, it becomes evident that achieving precise outcomes will require a deeper understanding of the anisotropy profiles.

References

Cao, S., Li, X., Biesiada, M., *et al.* 2017, *The Astrophysical Journal*, 835, 92
Collett, T. E., Oldham, L. J., Smith, R. J., *et al.* 2018, *Science*, 360, 1342
Koopmans, L. V. E., Bolton, A., Treu, T., *et al.* 2009, *The Astrophysical Journal*, 703, L51
Wang, Y., Vogelsberger, M., Xu, D., *et al.* 2019, *Mon. Not. R. Astron. Soc.*, 491, 5188
Osipkov, L. P. 1979, *Pisma v Astronomicheskii Zhurnal*, 5, 77
Merritt, D. 1985, *Mon. Not. R. Astron. Soc.*, 214, 25P
Mamon, G. A., Łokas, E. L. 2005, *Mon. Not. R. Astron. Soc.*, 363, 705

Chapter 5

Strong Gravitational Lensing in the Era of Big Data
Proceedings IAU Symposium No. 381, 2024
H. Stacey, A. Sonnenfeld & C. Grillo, eds.
doi:10.1017/S1743921323003873

Learning about the structure of strongly lensed AGNs from their lightcurves

Dominique Sluse

University of Liège
STAR Institute, Quartier Agora - Allée du six Août, 19c B-4000 Liège, Belgium.
email: dsluse@uliege.be

Abstract. The lightcurves of strongly lensed AGNs get distorted due to gravitational microlensing, which differently magnifies the emission regions of AGNs depending on their size. This effect has been used to measure the size of the AGN accretion disc, but high photometric accuracy lightcurves reveal coherent variations on short time-scales that are not expected by standard accretion disc models. I show that this signal can be produced by emission from the Broad Line Region (BLR) but also by extended (diffuse) continuum emission. I explain how these features can be used to measure the size of the BLR but also reveal additional sources of emission. The multi-colour lightcurves of lensed AGNs, such as those to be obtained with the Vera Rubin Observatory, may become a powerful new tool to reveal the sub-parsec structure of AGNs, and shed light on elusive AGN emitting regions such as the one producing diffuse continuum emission.

Keywords. gravitational lensing, micro-lensing

1. Microlensing and AGN structure

The multiple images of strongly lensed active galactic nuclei (AGN) constitute powerful natural laboratories for studying the physical mechanisms at work in the direct vicinity of distant supermassive black holes. Strong lensing provides natural magnification which turns lensing galaxies into natural telescopes, giving access to fainter objects for a fixed observational setup. Because of the time delay existing between lensed images, strong lensing even works as a time-machine, providing images of the same system up to hundreds of days before/after their actual observations. In general the individual lightcurves of each lensed image do not perfectly match after correcting for the time-delay. The reason is the microlensing produced by the stars in the lensing galaxy. While the microlensing signal has been seen as a nuisance for a number of applications, such as time-delay cosmography (e.g. Tewes et al. 2013), it encodes information about the structure of the lensed AGNs on the scale of a stellar-mass Einstein radius, namely an angular scale of micro-arcseconds (i.e. $< 10^{-2}$ pc in the source plane). Since the phenomenon is dynamic, with substantial variations of microlensing amplitude taking place on timescales ranging from a few months to several years, tracking microlensing variability enables one to virtually "scan" the inner-most part of AGNs otherwise unresolved with existing observational facilities.

Because the dominant contribution to AGN emission in the near-UV/optical range arises from the power-law continuum emission, it means that most of the microlensing signal detected in optical lightcurves is caused by microlensing of the accretion disc producing that continuum. The amplitude of the microlensing signal directly scales with the size of source: the smaller the source is compared to the (microlensing) Einstein radius, the stronger is the expected strength of the microlensing signal. This property has been used to measure the size of the accretion disc, commonly found to be larger than the prediction of the standard Shakura-Sunyaev accretion disc model. Moreover, because the disc size is more compact in the blue than in the red, the amplitude of microlensing may in general decrease with increasing wavelength. Consequently, by studying the chromatic changes of microlensing as a function of time, it becomes possible to constrain the wavelength dependence of the effective size of the disc. This method nicely complements continuum reverberation mapping of the disc, which requires very high photometric accuracy and a dense observational sampling. Compared to reverberation mapping, microlensing shines up for more distant AGNs (as strong lensing is required) and accommodates lower sampling of the light curves. For those reasons, the large synoptic survey telescope (LSST) is expected to be a fantastic tool for constraining the accretion discs of hundreds of distant lensed quasars through the analysis of their multicolor lightcurves (Neira et al. 2020).

The existing forecasts of microlensing signal as well as models of observed data display smooth (roller-coaster like) variations of the magnification as a function of time. While this effectively reproduces the main variations seen in microlensing lightcurves (obtained by taking the magnitude difference between lightcurves of pairs of lensed images after correcting for the time-delay), one often detects additional flickering beyond the noise level. This additional flickering, which appears like faster and smaller amplitude microlensing, has been identified in data already at the end of the nineties (e.g. Schild 1996; Gould & Miralda-Escudé 1997; Schechter et al. 2003), but it has received moderate attention since then. The lightucurves gathered by the COSMOGRAIL collaboration reveal that this signal is genuine, and present in many systems (e.g. Millon et al. 2020). Figure 1 shows an example of the "fast apparent microlensing" features seen in RXS J1131-1231.

The explanations for this signal fall in two categories: (a) it is caused by an unaccounted low mass (e.g. planetary-like) population of microlenses; (b) it is due to unaccounted structures in the source. In this work, we focus on the second category of explanations. In particular, we notice an important simplification when interpreting broad-band fluxes, which is that the whole flux is supposed to arise from the accretion disc. This assumption is however in general incorrect as flux from the broad emission lines is superimposed to the power-law continuum emission. This line flux often represents 30-40% of the flux in a given band, especially when iron emission is significant. Additional continuum emission arising from regions beyond the disc may not be excluded either but its reality is still debated.

In the following, we explore how emission arising from regions more extended than the continuum, and hence much less microlensed, modify the difference lightcurve between pairs of lensed images after time-shift for time-delay (hereafter generally called microlensing lightcurves). We discuss two scenarios: in the first one, we consider emission from the BLR, as originally proposed by Sluse & Tewes (2014); in the second one, we consider diffuse continuum emission (in addition to the accretion disc flux) arising from a region larger than the BLR. For both situations, we assume that the extended region reverberates the flux from the disc, with some time-lag $\tau = d/c$, where c is the speed of light and d is the distance between the continuum and the extended emission.

2. Method and main results

Our toy model emulates microlensed broad band lightcurves of AGNs with two emitting regions: a compact emission $F_{\rm cont}(t)$ that arises exclusively from the disc and which is sufficiently compact to be microlensed, and a second component, too large to be microlensed, which responds to the continuum after a time τ. For simplicity, we assume that this extended region responds instantly to the incident continuum flux after a time τ. This

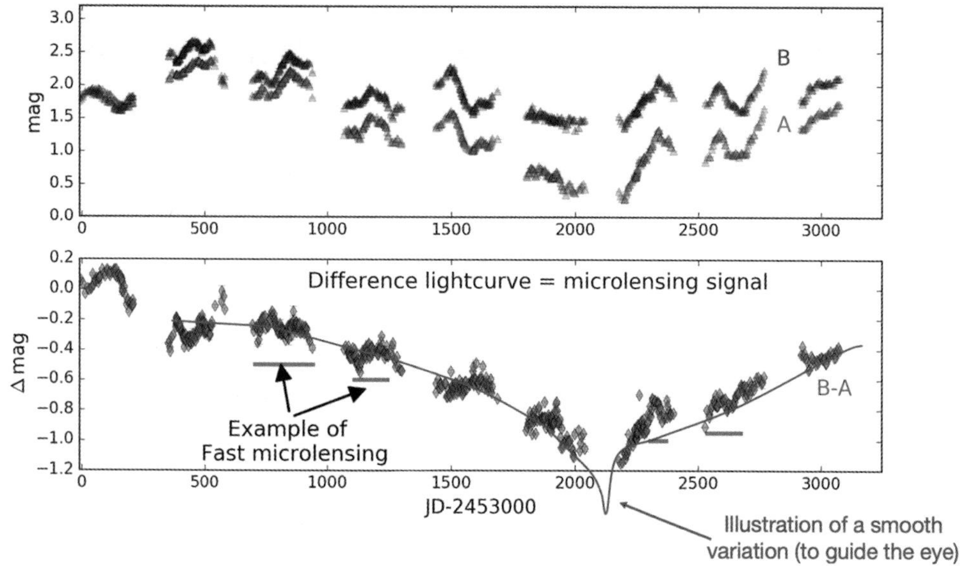

Figure 1. *Top:* Lightcurves of images A & B of the lensed AGN J1131-1231 obtained by Tewes et al. (2013). *Bottom:* Difference lightcurve enabling to cancel the intrinsic variability of the AGN and isolate the microlensing signal. The blue curve shows the signal (qualitative trend) that would be expected in case of microlensing of an accretion disc. The horizontal red lines indicate small amplitude "fast" variations arising on time scales of hundreds of days. Those features may be caused by structures in the source absent of existing AGN models.

academic assumption is equivalent to say that the transfer function of the reprocessing region is a Dirac $\delta(t)$ function. A more realistic situation would consists in convolving the input signal with a transfer function $\Psi(t)$, but the latter is a priori unknown. It would generally smooth out the reverberated flux. While this may have a qualitative impact on the shape of the discussed features, the general results discussed below will remain. The intrinsic variability signal $F_{\rm cont}(t)$ of the AGN is generated using a damped random walk model, and an arbitrary microlensing magnification $\mu(t)$. To ease the interpretation, we apply microlensing to only one of the lensed images, which is equivalent to fix $\mu(t)=1$ for the second image. The intrinsic variability signal as well as the microlensing factors are chosen to qualitatively reproduce the data of J1131-1231 displayed in Fig. 1 (Sluse et al., in prep). Ignoring the multiplicative macro-magnification of the image, we can express the observed flux F_i of the lensed image i as:

$$F_i = (1-f) \times \mu(t) \times F_{\rm cont}(t) + f \times F^*_{\rm cont}(t+\tau), \tag{1}$$

where f is the fraction of flux arising from a region larger than the continuum (we assume a single region), and $F^*_{\rm cont}(t+\tau)$ is the reverberated flux.

In our simulation, we have varied f in a range [0.1, 0.5], and consider two sets of values for τ. On the one hand, we have looked at "short time-lags" with $\tau \in [10, 50]$ days, which corresponds to size of the region emitting $H\beta$ for the lens system J1131−1231. Second, we have considered the case of time lags an order of magnitude larger, namely in the range 100-500 days. This second situation corresponds to an hypothetic diffuse continuum emission emitted in the outskirts of the BLR. The details of those results will be presented in Sluse et al. (in prep) but we outline here some of our most salient findings.

Figure 2 displays the result of a simulation where $f=0.5$ and the cases of a BLR-type lag (blue) and of a diffuse-continuum-type lag (500 days). The key features we observe are (a) the true micro-magnification is biased "proportionally" the fraction of

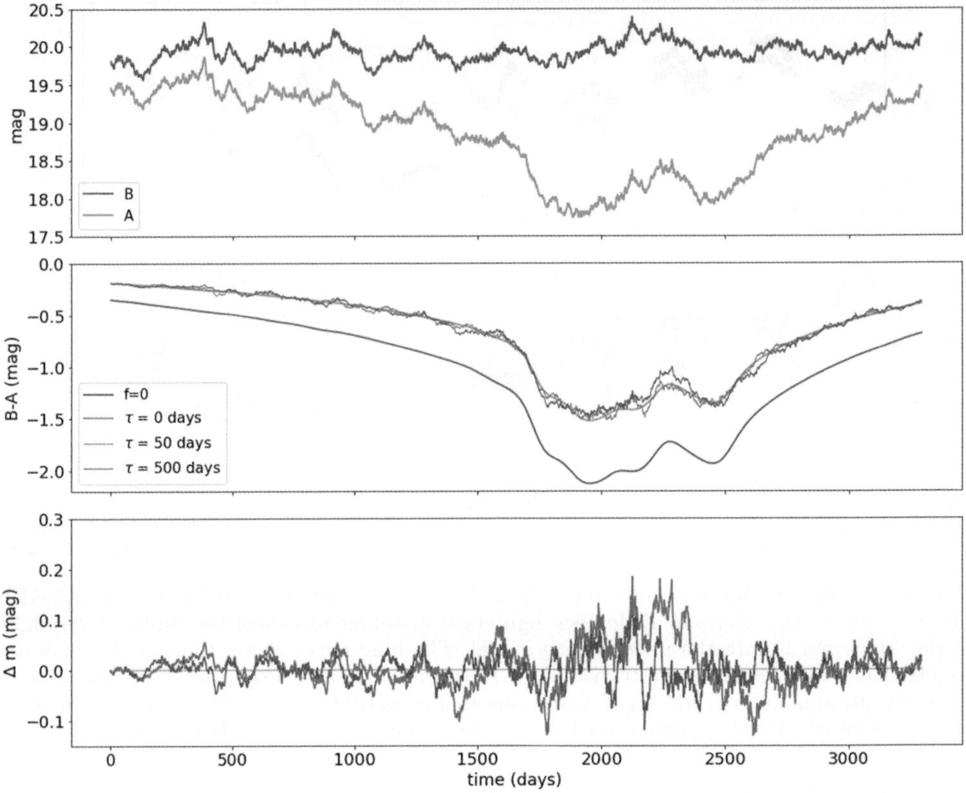

Figure 2. *Top:* Simulated broad band lighturves of a pair of images according to Eq. 1. The panel displays a situation where $f = 0.5$, and only image A is affected by a microlensing magnification $\mu(t)$ (for an accretion-disc-like source) shown in blue in the middle panel. Fluxes have been converted to magnitude. *Middle:* Difference lightcurve $m_B - m_A$. The blue curve is for $f = 0$, while the other curves show the case where $f = 0.5$ and $\tau = 50$ days (green, BLR case), and $\tau = 500$ days (red, diffuse continuum). The orange lightcurve shows the "academic" situation where $\tau = 0$. *Bottom:* Difference between the reference ($\tau = 0$ day) and lightcurves with $\tau = 50$ days (green) and $\tau = 500$ days (red). The largest flickering is obtained when the microlensing amplitude reaches a maximum ($t \sim$ 2000–2500 days) and $\tau = 500$ days.

non microlensed flux (the bias approximately scales with $2.5 \log(1 - f)$). (b) the amplitude of the fluctuations depends on the amplitude of intrinsic variability and of f: the larger is f, the larger is the observed fluctuation. (c) the amplitude of the fluctuation is minimum when $\mu(t)$ gets closer to 1; (d) for a fixed f, the amplitude of the fluctuation is larger for larger τ, with variation reflecting the time scale of the intrinsic variability possible over the period of the lag.

At the light of these results, we see that a flickering around the smoothly varying microlensing with an amplitude of about ± 0.1 mag may happen in our toy model for $f = 0.5$ and a time-lag larger than typically 200 days (only the 500 days lag is shown). This is in qualitative agreement with what is seen in the ligthcurve of J1131−1231. As the emission from the BLR (mostly FeII) contributes only 30% of the R-band flux, it cannot explain the observed large amplitude signal. This suggests that the "fast" microlensing observed in that system could be caused by "continuum" emission arising from the outer part of the BLR or beyond. A set of spectroscopic monitoring data, not shown here, supports that finding (Sluse et al. in prep.).

3. Discussion and perspectives

There is increasing observational support for the existence of extended continuum emission in addition to the standard emission from the accretion disc, but its exact origin is yet debated (e.g. Lawrence 2012; Netzer 2022). For instance, Sluse et al. (2015); Hutsemékers et al. (2015, 2020), have shown, based on the analysis of the spectra of two Broad Absorption Line quasars, that about 40% of the UV continuum emission arises in those systems from a region more extended than the accretion disc. Diffuse continuum emission in AGNs is suspected since several decades (e.g. Korista et al. 1995; Korista & Goad 2001) but its unambiguous detection is observationally challenging. The role of diffuse emission in interpreting continuum reverberation mapping data, which are in general hardly reconciled with a Shakura-Sunyaev accretion disc model, is not completely settled (e.g. Chelouche et al. 2019; Cackett et al. 2020; Guo et al. 2022). The data from microlensed quasars as well as the simulations presented in this work show that the detailed study of microlensed AGNs may provide important clues on the existence and properties of extended continuum. High accuracy photometric monitoring of strongly lensed quasars therefore provide a new technique to study the elusive extended continuum emission present in AGNs.

The advent of LSST, providing multi-colour lightcurves for an ensemble of lensed quasars, may enable to probe the nature of this emission, offering important insights on the "mechanics" of AGNs. Several observational signatures may help identifying and characterizing the properties of any diffuse continuum emission:

• Since the fraction of BLR flux in each band can be estimated from spectroscopy, it may be possible to compare the observed amplitude of flickering to model expectations. Deviations from this prediction may indicate additional amount of continuum emission.

• Multiple lensed images provide complementary information: images strongly demagnified due to microlensing may reveal flux variations dominated by the extended continuum while large micro-magnification events will display flux dominated by the disc emission.

• The idea proposed in Sluse & Tewes (2014) of doing photometric reverberation mapping by using lightcurves of multiple lensed images can be applied to the study of diffuse continuum, but deblending the emission of any extended continuum from real BLR flux may require anciallary spectroscopic data. The use of a power spectrum analysis of the microlensed signal may also be a promising alternative to standard reverberation mapping that is based on cross-correlation between signals (Paic et al. 2022).

The addition of multi-epoch spectroscopy (even sparse) may help to break model degeneracies from multi-band LSST data (e.g. by fixing the fraction of flux from the BLR at specific epochs). Facilities such as the Wide Field Spectroscopic telescope (Ellis et al. 2017) would a great asset for this kind of work.

The promises of using microlensing to study the properties of strongly lensed AGNs may be even greater than anticipated. The analysis of hundreds of microlensed systems may enable to simultaneously probe, the properties of the accretion disc, of the BLR, but also of an overlooked ingredient: diffuse continuum emission. A revision of our detailed understanding of AGNs is on its way.

References

Cackett, E. M., Gelbord, J., Li, Y.-R., et al. 2020, ApJ, 896, 1
Chelouche, D., Pozo Nuñez, F., & Kaspi, S. 2019, Nature Astronomy, 3, 251
Ellis, R. S., Bland-Hawthorn, J., Bremer, M., et al. 2017, arXiv e-prints, arXiv:1701.01976
Gould, A. & Miralda-Escudé, J. 1997, ApJl, 483, L13
Guo, H., Barth, A. J., & Wang, S. 2022, ApJ, 940, 20
Hutsemékers, D., Sluse, D., Braibant, L., & Anguita, T. 2015, A&A, 584, A61

Hutsemékers, D., Sluse, D., & Kumar, P. 2020, A&A, 633, A101
Korista, K. T., Alloin, D., Barr, P., et al. 1995, ApJs, 97, 285
Korista, K. T. & Goad, M. R. 2001, ApJ, 553, 695
Lawrence, A. 2012, MNRAS, 423, 451
Millon, M., Courbin, F., Bonvin, V., et al. 2020, A&A, 640, A105
Neira, F., Anguita, T., & Vernardos, G. 2020, MNRAS, 495, 544
Netzer, H. 2022, MNRAS, 509, 2637
Paic, E., Vernardos, G., Sluse, D., et al. 2022, A&A, 659, A21
Schechter, P. L., Udalski, A., Szymański, M., et al. 2003, ApJ, 584, 657
Schild, R. E. 1996, ApJ, 464, 125
Sluse, D. & Tewes, M. 2014, A&A, 571, A60
Sluse, D., Hutsemékers, D., Anguita, T., Braibant, L., & Riaud, P. 2015, A&A, 582, A109
Tewes, M., Courbin, F., Meylan, G., et al. 2013, A&A, 556, A22

Strong Gravitational Lensing in the Era of Big Data
Proceedings IAU Symposium No. 381, 2024
H. Stacey, A. Sonnenfeld & C. Grillo, eds.
doi:10.1017/S174392132300368X

Where are the Eddington-limited starbursts? Gravitational lensing provides a way forward for sub-kiloparsec views of star formation

Patrick S. Kamieneski

Arizona State University
email: pkamiene@asu.edu

Abstract. In the past decade, submillimeter surveys have been employed to define samples of gravitationally-lensed dusty star-forming galaxies (DSFGs) at $z \sim 1-4$. These extreme objects ($L_{\rm IR} = 10^{12} - 10^{13.5}$ $L_\odot$) appear to form stars prodigiously at rates of $100 - 3000$ $M_\odot$ yr^{-1}. Using all-sky *Planck* and *WISE* surveys, and wide-area *Herschel* surveys, we have identified the PASSAGES sample, with some of the rarest hyper-luminous IR galaxies ever discovered. We have found that their globally-averaged star formation surface densities are always sub-Eddington, typically by an order of magnitude. This may suggest that our understanding of how radiation pressure from massive stars disrupts the collapse of molecular clouds (thereby quenching star formation) is flawed—or simply that smaller physical resolutions are necessary. With the aid of lensing, we can now capture the source-plane distribution of star formation at $\sim$100pc scales, letting us identify isolated super-Eddington regions where quenching is occurring.

Keywords. Strong gravitational lensing (1643), Starburst galaxies (1570), Ultraluminous infrared galaxies (1735)

1. Introduction

Stellar feedback—the deposition of momentum and energy into the interstellar medium (ISM) by the formation of stars—has come to be accepted as a fundamental process governing the evolution of a galaxy. In particular, this feedback appears to be responsible for preventing gas-rich galaxies from rapidly converting gas into stars at a rate consistent with the free-fall time (see review McKee and Ostriker 2007). This feedback has impacts on global scales of a galaxy, but is ultimately a more local process. The efficiency of star formation is low on galaxy-wide scales; defined as $\varepsilon_{\rm ff} \equiv \tau_{\rm ff} \cdot {\rm SFR}/M_{\rm gas}$—with star-formation rate SFR, total gas mass $M_{\rm gas}$, and free-fall time $\tau_{\rm ff}$—typical values are of order 1% (e.g. Kennicutt 1998; Krumholz and McKee 2005; Krumholz et al. 2012; Padoan and Nordlund 2011; Federrath and Klessen 2012; Utomo et al. 2018). Yet, the efficiency for giant molecular clouds (GMCs) themselves can be much higher, even approaching unity in some instances (e.g., Murray 2011). These observational constraints to test models of star formation have been difficult to derive, so it remains unclear what physical processes dominate in regulating the collapse of molecular gas.

Scoville (2003), Murray et al. (2005), and Thompson et al. (2005) first developed theoretical models by which radiation pressure in dusty environments could become the predominant source of stellar feedback. In particular, where the ISM is optically thick to IR emission (resulting from the reprocessing of UV photons from massive stars), the momentum of the photons against dust grains coupled to the gas becomes an efficient support against the self-gravity of molecular clouds. Such an environment is expected for objects like ultra-luminous infrared galaxies (ULIRGs; Lonsdale et al. 2006)—or more generally the population of dusty star-forming galaxies (DSFGs)—which host extreme dust-enshrouded starbursts. This Eddington-like limit at which radiation pressure exceeds self-gravity is estimated to occur at star formation rate surface densities of around $\Sigma_{\rm SFR} \equiv {\rm SFR}/(2\pi R_{\rm eff}^2) \sim 1000\ M_\odot\ {\rm yr}^{-1}\ {\rm kpc}^{-2}$ (Andrews and Thompson 2011; Simpson et al. 2015). However, as this limit is for globally-averaged $\Sigma_{\rm SFR}$, galaxies forming stars at overall sub-Eddington densities can still host super-Eddington clumps of star formation. Moreover, the measurement of effective radii $R_{\rm eff}$ for high-z (often compact) galaxies requires finer angular resolution than local ULIRGs. For these reasons, the application of gravitational lensing as a "cosmic telescope" is required to make progress in testing models of stellar feedback.

2. Sample Definition and Data

For this analysis, we employ the lensed DSFGs selected with *Planck* as part of the PASSAGES sample (*Planck* All-Sky Survey to Analyze Gravitationally-lensed Extreme Starbursts; Harrington et al. 2016, 2021; Berman et al. 2022; lens models from Kamieneski et al. 2023). We also make some comparisons to similar samples defined with the South Pole Telescope (SPT; e.g., Weiss et al. 2013; lens models from Spilker et al. 2016) and *Herschel* (e.g., Negrello et al. 2010, 2017; Wardlow et al. 2013; lens models from Bussmann et al. 2013, 2015). As the *Planck*-selected objects occupy the upper echelon of apparent infrared luminosities for their redshift range of $z \sim 1 - 3.5$ (see also objects identified independently by Cañameras et al. 2015), they offer perhaps the ideal view of extreme starburst events. Moreover, radiative transfer modeling of multi-transition molecular gas lines for the PASSAGES sample by Harrington et al. (2021) suggested that some objects would approach this limit, based on the ratio of their $L_{\rm IR}$ to ISM mass (relative to the threshold of 500 $L_\odot/M_\odot$; see Scoville 2003). Since the total flux-weighted magnifications for extended objects (i.e., starbursts rather than quasars) generally tend not to exceed $\mu \sim 10 - 20$ (e.g., Hezaveh et al. 2012), their large apparent luminosities were perhaps more easily explained by larger intrinsic luminosities rather than larger magnifications. As such, they are among the best candidates for maximal starbursts forming stars at surface densities close to (or above) the Eddington limit.

Apparent star formation rates were derived from their total infrared luminosities (Harrington et al. 2016; Berman et al. 2022), which captures the UV emission from massive stars that is reprocessed into the IR by dust. As this covers a wide wavelength regime, for which the spatial distribution of continuum can vary, it is not entirely trivial to characterize the size of the dust-emitting region. For our purpose, we use 1.1mm continuum imaging (rest-frame $\sim 250 - 500\mu$m) on the Rayleigh-Jeans tail. Band 6 ($1.1 - 1.4$ mm) observations were taken with the Atacama Large Millimeter/submillimeter Array as part of Cycle 5 program 2017.1.01214.S (PI: M. Yun), reaching synthesized beam resolutions of $0.4 - 0.8''$ and sensitivities of $0.07 - 0.33$ mJy. The PASSAGES sub-sample that has been observed with ALMA and has a credible strong lens model consists of 13 objects. Further details on the observations are provided in Berman et al. (2022) and Kamieneski et al. (2023).

3. Results and Discussion

Fortunately, the measurement of $\Sigma_{\rm SFR}$ is relatively insensitive to statistical and systematic uncertainties in the lens model, as it is connected to surface brightness, which is conserved by gravitational lensing. For example, if a lens model erroneously overestimates a galaxy size by a factor of k relative to the "true" value, then the magnification measured from the ratio of image-plane to source-plane area will be underestimated by a factor of $\sim k^2$; this then leads to factor of k^2 overestimate of the intrinsic luminosity $L_{\rm IR}$. As a result of this, both the numerator and denominator of $\Sigma_{\rm IR} = L_{\rm IR}/(2\pi R_{\rm eff}^2)$ are overestimated by k^2, and the effect essentially cancels out. Considered slightly differently, $\Sigma_{\rm IR}$ and $\Sigma_{\rm SFR}$ could even be measured directly from the image-plane, lensing-uncorrected data, were it not for the challenge that shear distorts the shape of the galaxy and makes the determination of effective radius more complicated. As a more direct comparison with unlensed DSFGs is desirable, we carry out lens modeling in order to estimate the global surface densities of star formation.

Full details of our lens modeling approach for the PASSAGES sample are presented in Kamieneski et al. (2023), but we summarize it in brief here. Only the foreground lensing mass is parameterized (typically with singular isothermal ellipsoid profiles, Kormann et al. 1994), with constraints set by only the locations of multiply-imaged components. These image families are determined using multi-wavelength information from *Hubble* Space Telescope (HST) at 1.6 μm, Gemini-S Observatory at r' and z' bands, ALMA at 1.1mm, and the *Karl G. Jansky* Very Large Array at 6 GHz. Models are optimized through Markov Chain Monte Carlo with LENSTOOL (Kneib et al. 1996; Jullo et al. 2007; Jullo and Kneib 2009), effectively minimizing the RMS of image-plane separations between observed vs. predicted image locations. Magnifications (and associated uncertainties) are estimated by randomly sampling from the multi-dimensional posterior distribution, reconstructing in the source plane by ray-tracing, and computing the ratio of image-plane to source-plane areas. Objects are reconstructed in the source plane without parameterization or regularization, as these extreme starbursts may be clumpy and highly irregular. A similar approach is taken for measuring intrinsic effective radii and uncertainties (see also Section 3 of Kamieneski et al. 2023).

Figure 1 shows the globally-averaged $\Sigma_{\rm SFR}$ vs. $R_{\rm eff}$ of the dust continuum for three samples of lensed DSFGs, with intrinsic values derived with lens models: PASSAGES (this work, and see Table 3 from Kamieneski et al. 2023), SPT (Spilker et al. 2016), and *Herschel* (Bussmann et al. 2013). Additionally, unlensed DSFGs from the ALESS sample with resolved size measurements from Hodge et al. (2019) are shown. We find that very few of the objects shown approach the theoretical Eddington limit (with none surpassing it), and the median value of the four samples ($\Sigma_{\rm SFR} \sim 58\ M_\odot\ {\rm yr}^{-1}\ {\rm kpc}^{-2}$, $R_{\rm eff} \sim 1.6$ kpc) is more than an order of magnitude below the threshold. This is in apparent contrast to local ultra-luminous infrared galaxies (ULIRGs; $L_{\rm IR} = 10^{12-13}\ L_\odot$), which have been found to frequently exceed the Eddington limit, even by over an order of magnitude in some cases (e.g., Barcos-Muñoz et al. 2017; Pereira-Santaella et al. 2021). The canonical and well-studied ULIRG Arp 220, for example, has regions forming stars at surface densities of $\Sigma_{\rm SFR} \sim 10^{4.1\pm0.1}\ M_\odot\ {\rm yr}^{-1}\ {\rm kpc}^{-2}$ (Barcos-Muñoz et al. 2015). On the other hand, Song et al. (2022) found predominantly sub-Eddington star formation in a broader sample of local luminous infrared galaxies (LIRGs; $L_{\rm IR} = 10^{11-12}\ L_\odot$) and ULIRGs. If the extreme hyper-luminous IR galaxies (HyLIRGs; $L_{\rm IR} > 10^{13}\ L_\odot$) seen at high-$z$ like the PASSAGES objects are in fact sub-Eddington, it might suggest simply that their order-of-magnitude increase in IR luminosity over ULIRGs is accompanied by a comparable ~ 0.5 dex increase in radius. Indeed, for a very large sample of 1000+ DSFGs covering LIRGs, ULIRGs, and a handful of HyLIRGs from $z = 0 - 6$, Fujimoto et al. (2017) found a highly significant correlation between $R_{\rm eff}$ and $L_{\rm IR}$, fit approximately by $R_{\rm eff} \propto L_{\rm IR}^{0.28\pm0.07}$ (albeit with large scatter). The authors speculate that the power-law slope could be set by processes of disk formation, given the similarity in slope for UV-detected star-forming galaxies (e.g., $\alpha_{\rm UV} = 0.27 \pm 0.01$, Shibuya et al. 2015). While perhaps a too simplistic picture for complicated high-z systems, it is useful to consider that the Stefan-Boltzmann law suggests that an optically-thick spherical blackbody should radiate as $L_{\rm IR} \propto R_{\rm eff}^2 T_{\rm dust}^4$.

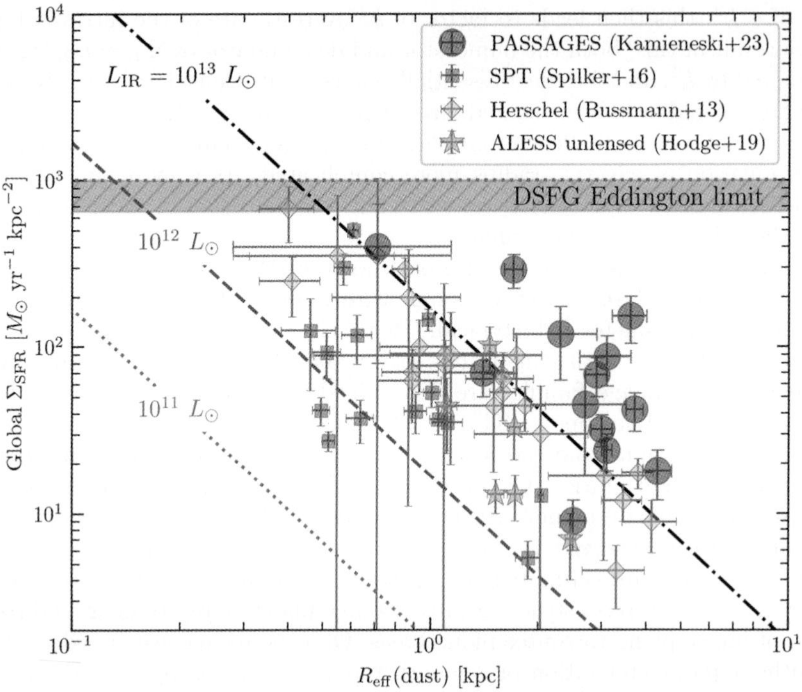

Figure 1. The galaxy-averaged star formation rate surface density Σ_{SFR} vs. dust continuum effective radius R_{eff} for members of the PASSAGES sample (Kamieneski et al. 2023), as measured in the model-reconstructed source plane. Also shown are the source-plane values for samples of lensed DSFGs identified with SPT (Spilker et al. 2016) and *Herschel* (Bussmann et al. 2013), alongside unlensed DSFGs from the ALESS sample (Hodge et al. 2019). The Eddington limit is typically computed at values of $\Sigma_{SFR} \sim 650 - 1000\ M_\odot$ yr^{-1} kpc^{-2} (depending on choice of gas mass fraction and dust-to-gas ratio; Andrews and Thompson 2011; Simpson et al. 2015; Hodge et al. 2019), which is shown as a red shaded region. Diagonal dotted, dashed, and dash-dotted lines indicate the loci for IR luminosities of $10^{11}\ L_\odot$ (LIRGs), $10^{12}\ L_\odot$ (ULIRGs), and $10^{13}\ L_\odot$ (hyper-luminous infrared galaxies or HyLIRGs).

However, as noted by Simpson et al. (2015), "clumpy" distributions of star formation with isolated sub-galactic regions—e.g., star-forming complexes and individual GMCs—forming stars above the Eddington limit could still appear sub-Eddington in its galaxy-wide star formation rate surface density. For this reason, in future work, we intend to directly map Σ_{SFR} at $\sim 100 - 500$pc scales with newly-obtained $\theta \sim 0.1''$ resolution ALMA observations of 16 PASSAGES objects at 870μm and 1.1mm. Even with very high-resolution $\theta \sim 0.07''$ ALMA imaging of six ALESS DSFGs, reaching 500pc physical scales, Hodge et al. (2019) found that both the peak and the globally-averaged Σ_{SFR} values remained sub-Eddington. However, these values were actually consistent with those of local ULIRGs from Barcos-Muñoz et al. (2017) with comparable physical sizes, and only ULIRGs with smaller effective radii ($\lesssim 0.2$ kpc) were found to be super-Eddington. Hodge et al. suggested this as a sign that even smaller physical scales need to be probed to find super-Eddington star formation in the distant Universe. Indeed, reaching physical scales of $\sim 60 - 160$ pc with lensing, Cañameras et al. (2017) found local densities of $\sim 1000 - 2000\ M_\odot$ yr^{-1} kpc^{-2}. At present, these resolutions are only feasibly accessible with the longest-baseline configurations of facilities like ALMA, and with the aid of gravitational lensing. However, the move to 100 pc scales also brings additional challenges. For example, measurement of the properties of small star-forming clumps may be more affected by gravitational microlensing (e.g., Fian et al. 2021), or by millilensing from dark, low-mass substructures in the lensing galaxy or anywhere along the line-of-sight (e.g., Mao and Schneider 1998; Vegetti et al. 2012, 2014; Inoue 2016; Nierenberg et al. 2017, 2020). Such effects are negligible at coarser resolutions, but they should be incorporated into uncertainties when approaching GMC scales. Fortunately, the conservation of surface brightness from lensing may again work to our advantage when mapping $\Sigma_{\rm SFR}$, whereas the measurement of absolute properties like the mass and luminosity of clumps is made more difficult.

Finally, we consider the future of submillimeter-selected samples of gravitational lenses like PASSAGES, which offer valuable laboratories for examining extreme star formation at ultra-high resolutions. At present, the PASSAGES sample consists of ~ 30 lensed DSFGs (Harrington et al. 2016; Berman et al. 2022). Other samples from *Planck*—and from facilities like *Herschel*, SPT, Atacama Cosmology Telescope (ACT)—contain an additional several hundred such lenses (e.g., Negrello et al. 2010, 2017; Vieira et al. 2010; González-Nuevo et al. 2012; Wardlow et al. 2013; Weiss et al. 2013; Su et al. 2017; Bakx et al. 2020a,b; Gralla et al. 2020; Trombetti et al. 2021; Lammers et al. 2022). While predicted by Blain (1996) and Negrello et al. (2007), submillimeter flux selection of gravitational lenses has proven to be remarkably efficacious, with few contaminants.

Novel selection criteria and new facilities will continue to increase this sample of lensed submillimeter-bright objects in the coming decade. As one notable example, the TolTEC camera of the Large Millimeter Telescope will have unprecedented sensitive mapping capabilities at 1.1mm, 1.4mm, and 2.1mm simultaneously (Wilson et al. 2020). This longer-wavelength regime could particularly help in accessing higher-redshift ($z \gtrsim 4$) lensed DSFGs (see, e.g., Casey et al. 2021). With this growing sample of lensed, intrinsically extreme starbursts—alongside the anticipated large ensemble of more than 10^5 optically-selected lenses with new mapping facilities like *Euclid* (e.g. Laureijs et al. 2011; Serjeant 2014)—new frontiers are being opened up for testing theoretical frameworks for star formation and the dynamics of the interstellar medium. Strong gravitational lensing will continue to be a crucial tool in helping to bridge the gap in the physical resolutions that we can reach for the local vs. high-redshift Universe.

References

Andrews, B. H. & Thompson, T. A. 2011, Assessing Radiation Pressure as a Feedback Mechanism in Star-forming Galaxies. *ApJ*, 727(2), 97.

Bakx, T. J. L. C., Dannerbauer, H., Frayer, D., Eales, S. A., Pérez-Fournon, I., Cai, Z. Y., Clements, D. L., De Zotti, G., González-Nuevo, J., Ivison, R. J., Lapi, A., Michałowski, M. J., Negrello, M., Serjeant, S., Smith, M. W. L., Temi, P., Urquhart, S., & van der Werf, P. 2020,a IRAM 30-m-EMIR redshift search of z = 3-4 lensed dusty starbursts selected from the HerBS sample. *MNRAS*, 496a(2), 2372–2390.

Bakx, T. J. L. C., Eales, S., & Amvrosiadis, A. 2020,b A search for the lenses in the Herschel Bright Sources (HerBS) sample. *MNRAS*, 493b(3), 4276–4293.

Barcos-Muñoz, L., Leroy, A. K., Evans, A. S., Condon, J., Privon, G. C., Thompson, T. A., Armus, L., Díaz-Santos, T., Mazzarella, J. M., Meier, D. S., Momjian, E., Murphy, E. J., Ott, J., Sanders, D. B., Schinnerer, E., Stierwalt, S., Surace, J. A., & Walter, F. 2017, A 33 GHz Survey of Local Major Mergers: Estimating the Sizes of the Energetically Dominant Regions from High-resolution Measurements of the Radio Continuum. *ApJ*, 843(2), 117.

Barcos-Muñoz, L., Leroy, A. K., Evans, A. S., Privon, G. C., Armus, L., Condon, J., Mazzarella, J. M., Meier, D. S., Momjian, E., Murphy, E. J., Ott, J., Reichardt, A., Sakamoto, K., Sanders, D. B., Schinnerer, E., Stierwalt, S., Surace, J. A., Thompson, T. A., & Walter, F. 2015, High-resolution Radio Continuum Measurements of the Nuclear Disks of Arp 220. *ApJ*, 799(1), 10.

Berman, D. A., Yun, M. S., Harrington, K. C., Kamieneski, P., Lowenthal, J., Frye, B. L., Wang, Q. D., Wilson, G. W., Aretxaga, I., Chavez, M., Cybulski, R., De la Luz, V., Erickson, N., Ferrusca, D., Hughes, D. H., Montaña, A., Narayanan, G., Sánchez-Argüelles, D., Schloerb, F. P., Souccar, K., Terlevich, E., Terlevich, R., & Zavala, J. A. 2022, PASSAGES: the Large Millimeter Telescope and ALMA observations of extremely luminous high-redshift galaxies identified by the Planck. *MNRAS*, 515(3), 3911–3937.

Blain, A. W. 1996, Galaxy-galaxy gravitational lensing in the millimetre/submillimetre waveband. *MNRAS*, 283, 1340–1348.

Bussmann, R. S., Pérez-Fournon, I., Amber, S., Calanog, J., Gurwell, M. A., Dannerbauer, H., De Bernardis, F., Fu, H., Harris, A. I., Krips, M., Lapi, A., Maiolino, R., Omont, A., Riechers, D., Wardlow, J., Baker, A. J., Birkinshaw, M., Bock, J., Bourne, N., Clements, D. L., Cooray, A., De Zotti, G., Dunne, L., Dye, S., Eales, S., Farrah, D., Gavazzi, R., González Nuevo, J., Hopwood, R., Ibar, E., Ivison, R. J., Laporte, N., Maddox, S., Martínez-Navajas, P., Michalowski, M., Negrello, M., Oliver, S. J., Roseboom, I. G., Scott, D., Serjeant, S., Smith, A. J., Smith, M., Streblyanska, A., Valiante, E., van der Werf, P., Verma, A., Vieira, J. D., Wang, L., & Wilner, D. 2013, Gravitational Lens Models Based on Submillimeter Array Imaging of Herschel-selected Strongly Lensed Sub-millimeter Galaxies at z > 1.5. *ApJ*, 779, 25.

Bussmann, R. S., Riechers, D., Fialkov, A., Scudder, J., Hayward, C. C., Cowley, W. I., Bock, J., Calanog, J., Chapman, S. C., Cooray, A., De Bernardis, F., Farrah, D., Fu, H., Gavazzi, R., Hopwood, R., Ivison, R. J., Jarvis, M., Lacey, C., Loeb, A., Oliver, S. J., Pérez-Fournon, I., Rigopoulou, D., Roseboom, I. G., Scott, D., Smith, A. J., Vieira, J. D., Wang, L., & Wardlow, J. 2015, HerMES: ALMA Imaging of Herschel-selected Dusty Star-forming Galaxies. *ApJ*, 812, 43.

Cañameras, R., Nesvadba, N., Kneissl, R., Frye, B., Gavazzi, R., Koenig, S., Le Floc'h, E., Limousin, M., Oteo, I., & Scott, D. 2017, Planck's dusty GEMS. IV. Star formation and feedback in a maximum starburst at z = 3 seen at 60-pc resolution. *A&A*, 604, A117.

Cañameras, R., Nesvadba, N. P. H., Guery, D., McKenzie, T., König, S., Petitpas, G., Dole, H., Frye, B., Flores-Cacho, I., Montier, L., Negrello, M., Beelen, A., Boone, F., Dicken, D., Lagache, G., Le Floc'h, E., Altieri, B., Béthermin, M., Chary, R., de Zotti, G., Giard, M., Kneissl, R., Krips, M., Malhotra, S., Martinache, C., Omont, A., Pointecouteau, E., Puget, J.-L., Scott, D., Soucail, G., Valtchanov, I., Welikala, N., & Yan, L. 2015, Planck's dusty GEMS: The brightest gravitationally lensed galaxies discovered with the Planck all-sky survey. *A&A*, 581, A105.

Casey, C. M., Zavala, J. A., Manning, S. M., Aravena, M., Béthermin, M., Caputi, K. I., Champagne, J. B., Clements, D. L., Drew, P., Finkelstein, S. L., Fujimoto, S., Hayward, C. C., Dekel, A. M., Kokorev, V., Lagos, C. d. P., Long, A. S., Magdis, G. E., Man, A. W. S., Mitsuhashi, I., Popping, G., Spilker, J., Staguhn, J., Talia, M., Toft, S., Treister, E., Weaver, J. R., & Yun, M. 2021, Mapping Obscuration to Reionization with ALMA (MORA): 2 mm Efficiently Selects the Highest-redshift Obscured Galaxies. *ApJ*, 923(2), 215.

Federrath, C. & Klessen, R. S. 2012, The Star Formation Rate of Turbulent Magnetized Clouds: Comparing Theory, Simulations, and Observations. *ApJ*, 761(2), 156.

Fian, C., Mediavilla, E., Jiménez-Vicente, J., Motta, V., Muñoz, J. A., Chelouche, D., Goméz-Alvarez, P., Rojas, K., & Hanslmeier, A. 2021, Revealing the structure of the lensed quasar Q 0957+561. I. Accretion disk size. *A&A*, 654, A70.

Fujimoto, S., Ouchi, M., Shibuya, T., & Nagai, H. 2017, Demonstrating a New Census of Infrared Galaxies with ALMA (DANCING-ALMA). I. FIR Size and Luminosity Relation at z = 0-6 Revealed with 1034 ALMA Sources. *ApJ*, 850, 83.

González-Nuevo, J., Lapi, A., Fleuren, S., Bressan, S., Danese, L., De Zotti, G., Negrello, M., Cai, Z. Y., Fan, L., Sutherland, W., Baes, M., Baker, A. J., Clements, D. L., Cooray, A., Dannerbauer, H., Dunne, L., Dye, S., Eales, S., Frayer, D. T., Harris, A. I., Ivison, R., Jarvis, M. J., Michałowski, M. J., López-Caniego, M., Rodighiero, G., Rowlands, K., Serjeant, S., Scott, D., van der Werf, P., Auld, R., Buttiglione, S., Cava, A., Dariush, A.,

Fritz, J., Hopwood, R., Ibar, E., Maddox, S., Pascale, E., Pohlen, M., Rigby, E., Smith, D., & Temi, P. 2012, Herschel-ATLAS: Toward a Sample of ~1000 Strongly Lensed Galaxies. *ApJ*, 749(1), 65.

Gralla, M. B., Marriage, T. A., Addison, G., Baker, A. J., Bond, J. R., Crichton, D., Datta, R., Devlin, M. J., Dunkley, J., Dünner, R., Fowler, J., Gallardo, P. A., Hall, K., Halpern, M., Hasselfield, M., Hilton, M., Hincks, A. D., Huffenberger, K. M., Hughes, J. P., Kosowsky, A., López-Caraballo, C. H., Louis, T., Marsden, D., Moodley, K., Niemack, M. D., Page, L. A., Partridge, B., Rivera, J., Sievers, J. L., Staggs, S., Su, T., Swetz, D., & Wollack, E. J. 2020, Atacama Cosmology Telescope: Dusty Star-forming Galaxies and Active Galactic Nuclei in the Equatorial Survey. *ApJ*, 893(2), 104.

Harrington, K. C., Weiss, A., Yun, M. S., Magnelli, B., Sharon, C. E., Leung, T. K. D., Vishwas, A., Wang, Q. D., Frayer, D. T., Jiménez-Andrade, E. F., Liu, D., García, P., Romano-Díaz, E., Frye, B. L., Jarugula, S., Bădescu, T., Berman, D., Dannerbauer, H., Díaz-Sánchez, A., Grassitelli, L., Kamieneski, P., Kim, W. J., Kirkpatrick, A., Lowenthal, J. D., Messias, H., Puschnig, J., Stacey, G. J., Torne, P., & Bertoldi, F. 2021, Turbulent Gas in Lensed Planck-selected Starbursts at z $\sim$ 1-3.5. *ApJ*, 908(1), 95.

Harrington, K. C., Yun, M. S., Çybulski, R., Wilson, G. W., Aretxaga, I., Chavez, M., De la Luz, V., Erickson, N., Ferrusca, D., Gallup, A. D., Hughes, D. H., Montaña, A., Narayanan, G., Sánchez-Argüelles, D., Schloerb, F. P., Souccar, K., Terlevich, E., Terlevich, R., Zeballos, M., & Zavala, J. A. 2016, Early science with the Large Millimeter Telescope: observations of extremely luminous high-z sources identified by Planck. *MNRAS*, 458, 4383–4399.

Hezaveh, Y. D., Marrone, D. P., & Holder, G. P. 2012, Size Bias and Differential Lensing of Strongly Lensed, Dusty Galaxies Identified in Wide-Field Surveys. *ApJ*, 761, 20.

Hodge, J. A., Smail, I., Walter, F., da Cunha, E., Swinbank, A. M., Rybak, M., Venemans, B., Brandt, W. N., Calistro Rivera, G., Chapman, S. C., Chen, C.-C., Cox, P., Dannerbauer, H., Decarli, R., Greve, T. R., Knudsen, K. K., Menten, K. M., Schinnerer, E., Simpson, J. M., van der Werf, P., Wardlow, J. L., & Weiss, A. 2019, ALMA Reveals Potential Evidence for Spiral Arms, Bars, and Rings in High-redshift Submillimeter Galaxies. *ApJ*, 876, 130.

Inoue, K. T. 2016, On the origin of the flux ratio anomaly in quadruple lens systems. *MNRAS*, 461(1), 164–175.

Jullo, E. & Kneib, J.-P. 2009, Multiscale cluster lens mass mapping - I. Strong lensing modelling. *MNRAS*, 395, 1319–1332.

Jullo, E., Kneib, J.-P., Limousin, M., Elíasdóttir, Á., Marshall, P. J., & Verdugo, T. 2007, A Bayesian approach to strong lensing modelling of galaxy clusters. *New Journal of Physics*, 9, 447.

Kamieneski, P. S., Yun, M. S., Harrington, K. C., Lowenthal, J. D., Wang, Q. D., Frye, B. L., Jimenez-Andrade, E. F., Vishwas, A., Cooper, O., Pascale, M., Foo, N., Berman, D., Englert, A., & Garcia Diaz, C. 2023, PASSAGES: the wide-ranging, extreme intrinsic properties of Planck-selected, lensed dusty star-forming galaxies. *arXiv e-prints*, arXiv:2301.09746.

Kennicutt, Robert C., J. 1998, The Global Schmidt Law in Star-forming Galaxies. *ApJ*, 498(2), 541–552.

Kneib, J.-P., Ellis, R. S., Smail, I., Couch, W. J., & Sharples, R. M. 1996, Hubble Space Telescope Observations of the Lensing Cluster Abell 2218. *ApJ*, 471, 643.

Kormann, R., Schneider, P., & Bartelmann, M. 1994, Isothermal elliptical gravitational lens models. *A&A*, 284, 285–299.

Krumholz, M. R., Dekel, A., & McKee, C. F. 2012, A Universal, Local Star Formation Law in Galactic Clouds, nearby Galaxies, High-redshift Disks, and Starbursts. *ApJ*, 745(1), 69.

Krumholz, M. R. & McKee, C. F. 2005, A General Theory of Turbulence-regulated Star Formation, from Spirals to Ultraluminous Infrared Galaxies. *ApJ*, 630, 250–268.

Lammers, C., Hill, R., Lim, S., Scott, D., Cañameras, R., & Dole, H. 2022, Candidate high-redshift protoclusters and lensed galaxies in the Planck list of high-z sources overlapping with Herschel-SPIRE imaging. *MNRAS*, 514(4), 5004–5023.

Laureijs, R., Amiaux, J., Arduini, S., Augueres, J. L., Brinchmann, J., Cole, R., Cropper, M., Dabin, C., Duvet, L., Ealet, A., & et al. 2011, Euclid Definition Study Report. *arXiv e-prints*, arXiv:1110.3193.

Lonsdale, C. J., Farrah, D., & Smith, H. E. 2006, *Ultraluminous Infrared Galaxies*, 285. Springer-Verlag.

Mao, S. & Schneider, P. 1998, Evidence for substructure in lens galaxies? *MNRAS*, 295, 587.

McKee, C. F. & Ostriker, E. C. 2007, Theory of Star Formation. *ARA&A*, 45(1), 565–687.

Murray, N. 2011, Star Formation Efficiencies and Lifetimes of Giant Molecular Clouds in the Milky Way. *ApJ*, 729(2), 133.

Murray, N., Quataert, E., & Thompson, T. A. 2005, On the Maximum Luminosity of Galaxies and Their Central Black Holes: Feedback from Momentum-driven Winds. *ApJ*, 618(2), 569–585.

Negrello, M., Amber, S., Amvrosiadis, A., Cai, Z.-Y., Lapi, A., Gonzalez-Nuevo, J., De Zotti, G., Furlanetto, C., Maddox, S. J., Allen, M., Bakx, T., Bussmann, R. S., Cooray, A., Covone, G., Danese, L., Dannerbauer, H., Fu, H., Greenslade, J., Gurwell, M., Hopwood, R., Koopmans, L. V. E., Napolitano, N., Nayyeri, H., Omont, A., Petrillo, C. E., Riechers, D. A., Serjeant, S., Tortora, C., Valiante, E., Verdoes Kleijn, G., Vernardos, G., Wardlow, J. L., Baes, M., Baker, A. J., Bourne, N., Clements, D., Crawford, S. M., Dye, S., Dunne, L., Eales, S., Ivison, R. J., Marchetti, L., Michałowski, M. J., Smith, M. W. L., Vaccari, M., & van der Werf, P. 2017, The Herschel-ATLAS: a sample of 500 μm-selected lensed galaxies over 600 deg^2. *MNRAS*, 465, 3558–3580.

Negrello, M., Hopwood, R., De Zotti, G., Cooray, A., Verma, A., Bock, J., Frayer, D. T., Gurwell, M. A., Omont, A., Neri, R., Dannerbauer, H., Leeuw, L. L., Barton, E., Cooke, J., Kim, S., da Cunha, E., Rodighiero, G., Cox, P., Bonfield, D. G., Jarvis, M. J., Serjeant, S., Ivison, R. J., Dye, S., Aretxaga, I., Hughes, D. H., Ibar, E., Bertoldi, F., Valtchanov, I., Eales, S., Dunne, L., Driver, S. P., Auld, R., Buttiglione, S., Cava, A., Grady, C. A., Clements, D. L., Dariush, A., Fritz, J., Hill, D., Hornbeck, J. B., Kelvin, L., Lagache, G., Lopez-Caniego, M., Gonzalez-Nuevo, J., Maddox, S., Pascale, E., Pohlen, M., Rigby, E. E., Robotham, A., Simpson, C., Smith, D. J. B., Temi, P., Thompson, M. A., Woodgate, B. E., York, D. G., Aguirre, J. E., Beelen, A., Blain, A., Baker, A. J., Birkinshaw, M., Blundell, R., Bradford, C. M., Burgarella, D., Danese, L., Dunlop, J. S., Fleuren, S., Glenn, J., Harris, A. I., Kamenetzky, J., Lupu, R. E., Maddalena, R. J., Madore, B. F., Maloney, P. R., Matsuhara, H., Michałowski, M. J., Murphy, E. J., Naylor, B. J., Nguyen, H., Popescu, C., Rawlings, S., Rigopoulou, D., Scott, D., Scott, K. S., Seibert, M., Smail, I., Tuffs, R. J., Vieira, J. D., van der Werf, P. P., & Zmuidzinas, J. 2010, The Detection of a Population of Submillimeter-Bright, Strongly Lensed Galaxies. *Science*, 330, 800.

Negrello, M., Perrotta, F., González-Nuevo, J., Silva, L., de Zotti, G., Granato, G. L., Baccigalupi, C., & Danese, L. 2007, Astrophysical and cosmological information from large-scale submillimetre surveys of extragalactic sources. *MNRAS*, 377, 1557–1568.

Nierenberg, A. M., Gilman, D., Treu, T., Brammer, G., Birrer, S., Moustakas, L., Agnello, A., Anguita, T., Fassnacht, C. D., Motta, V., Peter, A. H. G., & Sluse, D. 2020, Double dark matter vision: twice the number of compact-source lenses with narrow-line lensing and the WFC3 grism. *MNRAS*, 492(4), 5314–5335.

Nierenberg, A. M., Treu, T., Brammer, G., Peter, A. H. G., Fassnacht, C. D., Keeton, C. R., Kochanek, C. S., Schmidt, K. B., Sluse, D., & Wright, S. A. 2017, Probing dark matter substructure in the gravitational lens HE 0435-1223 with the WFC3 grism. *MNRAS*, 471(2), 2224–2236.

Padoan, P. & Nordlund, Å. 2011, The Star Formation Rate of Supersonic Magnetohydrodynamic Turbulence. *ApJ*, 730(1), 40.

Pereira-Santaella, M., Colina, L., García-Burillo, S., Lamperti, I., González-Alfonso, E., Perna, M., Arribas, S., Alonso-Herrero, A., Aalto, S., Combes, F., Labiano, A., Piqueras-López, J., Rigopoulou, D., & van der Werf, P. 2021, Physics of ULIRGs with MUSE and ALMA: The PUMA project. II. Are local ULIRGs powered by AGN? The subkiloparsec view of the 220 GHz continuum. *A&A*, 651, A42.

Scoville, N. 2003, Starburst and AGN Connections and Models. *Journal of Korean Astronomical Society*, 36(3), 167–175.

Serjeant, S. 2014, Up to 100,000 Reliable Strong Gravitational Lenses in Future Dark Energy Experiments. *ApJL*, 793, L10.

Shibuya, T., Ouchi, M., & Harikane, Y. 2015, Morphologies of ∼190,000 Galaxies at z = 0-10 Revealed with HST Legacy Data. I. Size Evolution. *ApJS*, 219, 15.

Simpson, J. M., Smail, I., Swinbank, A. M., Almaini, O., Blain, A. W., Bremer, M. N., Chapman, S. C., Chen, C.-C., Conselice, C., Coppin, K. E. K., Danielson, A. L. R., Dunlop, J. S., Edge, A. C., Farrah, D., Geach, J. E., Hartley, W. G., Ivison, R. J., Karim, A., Lani, C., Ma, C. J., Meijerink, R., Michałowski, M. J., Mortlock, A., Scott, D., Simpson, C. J., Spaans, M., Thomson, A. P., van Kampen, E., & van der Werf, P. P. 2015, The SCUBA-2 Cosmology Legacy Survey: ALMA Resolves the Rest-frame Far-infrared Emission of Sub-millimeter Galaxies. *ApJ*, 799(1), 81.

Song, Y., Linden, S. T., Evans, A. S., Barcos-Muñoz, L., Murphy, E. J., Momjian, E., Díaz-Santos, T., Larson, K. L., Privon, G. C., Huang, X., Armus, L., Mazzarella, J. M., U, V., Inami, H., Charmandaris, V., Ricci, C., Emig, K. L., McKinney, J., Yoon, I., Kunneriath, D., Lai, T. S. Y., Rodas-Quito, E. E., Saravia, A., Gao, T., Meynardie, W., & Sanders, D. B. 2022, Characterizing Compact 15-33 GHz Radio Continuum Sources in Local U/LIRGs. *ApJ*, 940(1), 52.

Spilker, J. S., Marrone, D. P., Aravena, M., Béthermin, M., Bothwell, M. S., Carlstrom, J. E., Chapman, S. C., Crawford, T. M., de Breuck, C., Fassnacht, C. D., Gonzalez, A. H., Greve, T. R., Hezaveh, Y., Litke, K., Ma, J., Malkan, M., Rotermund, K. M., Strandet, M., Vieira, J. D., Weiss, A., & Welikala, N. 2016, ALMA Imaging and Gravitational Lens Models of South Pole Telescope – Selected Dusty, Star-Forming Galaxies at High Redshifts. *ApJ*, 826, 112.

Su, T., Marriage, T. A., Asboth, V., Baker, A. J., Bond, J. R., Crichton, D., Devlin, M. J., Dünner, R., Farrah, D., Frayer, D. T., Gralla, M. B., Hall, K., Halpern, M., Harris, A. I., Hilton, M., Hincks, A. D., Hughes, J. P., Niemack, M. D., Page, L. A., Partridge, B., Rivera, J., Scott, D., Sievers, J. L., Thornton, R. J., Viero, M. P., Wang, L., Wollack, E. J., & Zemcov, M. 2017, On the redshift distribution and physical properties of ACT-selected DSFGs. *MNRAS*, 464, 968–984.

Thompson, T. A., Quataert, E., & Murray, N. 2005, Radiation Pressure-supported Starburst Disks and Active Galactic Nucleus Fueling. *ApJ*, 630(1), 167–185.

Trombetti, T., Burigana, C., Bonato, M., Herranz, D., De Zotti, G., Negrello, M., Galluzzi, V., & Massardi, M. 2021, Search for candidate strongly lensed dusty galaxies in the Planck satellite catalogues. *A&A*, 653, A151.

Utomo, D., Sun, J., Leroy, A. K., Kruijssen, J. M. D., Schinnerer, E., Schruba, A., Bigiel, F., Blanc, G. A., Chevance, M., Emsellem, E., Herrera, C., Hygate, A. P. S., Kreckel, K., Ostriker, E. C., Pety, J., Querejeta, M., Rosolowsky, E., Sandstrom, K. M., & Usero, A. 2018, Star Formation Efficiency per Free-fall Time in nearby Galaxies. *ApJL*, 861(2), L18.

Vegetti, S., Koopmans, L. V. E., Auger, M. W., Treu, T., & Bolton, A. S. 2014, Inference of the cold dark matter substructure mass function at z = 0.2 using strong gravitational lenses. *MNRAS*, 442(3), 2017–2035.

Vegetti, S., Lagattuta, D. J., McKean, J. P., Auger, M. W., Fassnacht, C. D., & Koopmans, L. V. E. 2012, Gravitational detection of a low-mass dark satellite galaxy at cosmological distance. *Nature*, 481(7381), 341–343.

Vieira, J. D., Crawford, T. M., Switzer, E. R., Ade, P. A. R., Aird, K. A., Ashby, M. L. N., Benson, B. A., Bleem, L. E., Brodwin, M., Carlstrom, J. E., Chang, C. L., Cho, H. M., Crites, A. T., de Haan, T., Dobbs, M. A., Everett, W., George, E. M., Gladders, M., Hall, N. R., Halverson, N. W., High, F. W., Holder, G. P., Holzapfel, W. L., Hrubes, J. D., Joy, M., Keisler, R., Knox, L., Lee, A. T., Leitch, E. M., Lueker, M., Marrone, D. P., McIntyre, V., McMahon, J. J., Mehl, J., Meyer, S. S., Mohr, J. J., Montroy, T. E., Padin, S., Plagge, T., Pryke, C., Reichardt, C. L., Ruhl, J. E., Schaffer, K. K., Shaw, L., Shirokoff, E., Spieler, H. G., Stalder, B., Staniszewski, Z., Stark, A. A., Vanderlinde, K., Walsh, W.,

Williamson, R., Yang, Y., Zahn, O., & Zenteno, A. 2010, Extragalactic Millimeter-wave Sources in South Pole Telescope Survey Data: Source Counts, Catalog, and Statistics for an 87 Square-degree Field. *ApJ*, 719(1), 763–783.

Wardlow, J. L., Cooray, A., De Bernardis, F., Amblard, A., Arumugam, V., Aussel, H., Baker, A. J., Béthermin, M., Blundell, R., Bock, J., Boselli, A., Bridge, C., Buat, V., Burgarella, D., Bussmann, R. S., Cabrera-Lavers, A., Calanog, J., Carpenter, J. M., Casey, C. M., Castro-Rodríguez, N., Cava, A., Chanial, P., Chapin, E., Chapman, S. C., Clements, D. L., Conley, A., Cox, P., Dowell, C. D., Dye, S., Eales, S., Farrah, D., Ferrero, P., Franceschini, A., Frayer, D. T., Frazer, C., Fu, H., Gavazzi, R., Glenn, J., González Solares, E. A., Griffin, M., Gurwell, M. A., Harris, A. I., Hatziminaoglou, E., Hopwood, R., Hyde, A., Ibar, E., Ivison, R. J., Kim, S., Lagache, G., Levenson, L., Marchetti, L., Marsden, G., Martinez-Navajas, P., Negrello, M., Neri, R., Nguyen, H. T., O'Halloran, B., Oliver, S. J., Omont, A., Page, M. J., Panuzzo, P., Papageorgiou, A., Pearson, C. P., Pérez-Fournon, I., Pohlen, M., Riechers, D., Rigopoulou, D., Roseboom, I. G., Rowan-Robinson, M., Schulz, B., Scott, D., Scoville, N., Seymour, N., Shupe, D. L., Smith, A. J., Streblyanska, A., Strom, A., Symeonidis, M., Trichas, M., Vaccari, M., Vieira, J. D., Viero, M., Wang, L., Xu, C. K., Yan, L., & Zemcov, M. 2013, HerMES: Candidate Gravitationally Lensed Galaxies and Lensing Statistics at Submillimeter Wavelengths. *ApJ*, 762, 59.

Weiss, A., De Breuck, C., Marrone, D. P., Vieira, J. D., Aguirre, J. E., Aird, K. A., Aravena, M., Ashby, M. L. N., Bayliss, M., Benson, B. A., Béthermin, M., Biggs, A. D., Bleem, L. E., Bock, J. J., Bothwell, M., Bradford, C. M., Brodwin, M., Carlstrom, J. E., Chang, C. L., Chapman, S. C., Crawford, T. M., Crites, A. T., de Haan, T., Dobbs, M. A., Downes, T. P., Fassnacht, C. D., George, E. M., Gladders, M. D., Gonzalez, A. H., Greve, T. R., Halverson, N. W., Hezaveh, Y. D., High, F. W., Holder, G. P., Holzapfel, W. L., Hoover, S., Hrubes, J. D., Husband, K., Keisler, R., Lee, A. T., Leitch, E. M., Lueker, M., Luong-Van, D., Malkan, M., McIntyre, V., McMahon, J. J., Mehl, J., Menten, K. M., Meyer, S. S., Murphy, E. J., Padin, S., Plagge, T., Reichardt, C. L., Rest, A., Rosenman, M., Ruel, J., Ruhl, J. E., Schaffer, K. K., Shirokoff, E., Spilker, J. S., Stalder, B., Staniszewski, Z., Stark, A. A., Story, K., Vanderlinde, K., Welikala, N., & Williamson, R. 2013, ALMA Redshifts of Millimeter-selected Galaxies from the SPT Survey: The Redshift Distribution of Dusty Star-forming Galaxies. *ApJ*, 767, 88.

Wilson, G. W., Abi-Saad, S., Ade, P., Aretxaga, I., Austermann, J., Ban, Y., Bardin, J., Beall, J., Berthoud, M., Bryan, S., Bussan, J., Castillo, E., Chavez, M., Contente, R., DeNigris, N. S., Dober, B., Eiben, M., Ferrusca, D., Fissel, L., Gao, J., Golec, J. E., Golina, R., Gomez, A., Gordon, S., Gutermuth, R., Hilton, G., Hosseini, M., Hubmayr, J., Hughes, D., Kuczarski, S., Lee, D., Lunde, E., Ma, Z., Mani, H., Mauskopf, P., McCrackan, M., McKenney, C., McMahon, J., Novak, G., Pisano, G., Pope, A., Ralston, A., Rodriguez, I., Sánchez-Argüelles, D., Schloerb, F. P., Simon, S., Sinclair, A., Souccar, K., Torres Campos, A., Tucker, C., Ullom, J., Van Camp, E., Van Lanen, J., Velazquez, M., Vissers, M., Weeks, E., & Yun, M. S. The TolTEC camera: an overview of the instrument and in-lab testing results. In Zmuidzinas, J. & Gao, J.-R., editors, *Millimeter, Submillimeter, and Far-Infrared Detectors and Instrumentation for Astronomy X* 2020, volume 11453 of *Society of Photo-Optical Instrumentation Engineers (SPIE) Conference Series*, 1145302.

Strong Gravitational Lensing in the Era of Big Data
Proceedings IAU Symposium No. 381, 2024
H. Stacey, A. Sonnenfeld & C. Grillo, eds.
doi:10.1017/S1743921323003666

Stellar mass fraction and quasar accretion disk size in SDSS J1004+4112 from photometric follow-up

R. Forés-Toribio[1,2], J. A. Muñoz[1,2], C. Fian[1], J. Jiménez-Vicente[3,4] and E. Mediavilla[5,6]

[1]Departamento de Astronomía y Astrofísica, Universidad de Valencia, E-46100 Burjassot, Valencia, Spain. email: raquel.fores@uv.es

[2]Observatorio Astronómico, Universidad de Valencia, E-46980 Paterna, Valencia, Spain

[3]Departamento de Física Teórica y del Cosmos, Universidad de Granada, Campus de Fuentenueva, 18071 Granada, Spain

[4]Instituto Carlos I de Física Teórica y Computacional, Universidad de Granada, 18071 Granada, Spain

[5]Instituto de Astrofísica de Canarias, Vía Láctea S/N, La Laguna, E-38200, Tenerife, Spain

[6]Departamento de Astrofísica, Universidad de la Laguna, La Laguna, E-38200, Tenerife, Spain

Abstract. The gravitational lens SDSS J1004+4112 was the first discovered system where a background quasar is lensed by a galaxy cluster instead of a single galaxy. We use the 14.5-year r-band light curves together with the recently measured time delay of the fourth brightest quasar image (Muñoz et al. (2022)) and the mass model from Forés-Toribio et al. (2022) to study the microlensing effect in this system. We constrain the quasar accretion disk size to $R_{1/2} = 5.3^{+1.3}_{-0.7}\sqrt{M/0.3M_{\odot}}$ light-days at 2407Å in the restframe which is compatible with most previous estimates. We also infer the fraction of mass in stars at the positions of the quasar images: $\alpha_A = 0.058^{+0.024}_{-0.032}$, $\alpha_B = 0.048^{+0.032}_{-0.014}$, $\alpha_C = 0.018^{+0.015}_{-0.018}$ and $\alpha_D = 0.008^{+0.033}_{-0.008}$. The stellar fraction estimates are reasonable for intracluster medium although the stellar fractions at images A and B are slightly larger, suggesting the presence of a near undetected galaxy.

Keywords. Gravitational lensing: micro, Galaxies: clusters: intracluster medium, Accretion, accretion disks, quasars: individual: SDSS J1004+4112

1. Introduction

The first example of a quasar lensed by a galaxy cluster was SDSS J1004+4112 (Inada et al. (2003)). Since the deflector mass is larger than for the typical scenario of lensed quasars, the multiple images are formed far (∼15") from the brightest cluster galaxy. With this particular configuration, the light from the quasar images travels mainly through the intracluster medium and the impact of microlensing was expected to be small. However, soon after its discovery, microlensing variability was reported in the blue wing of the broad emission lines of image A (Richards et al. (2004); Gómez-Álvarez et al. (2006); Lamer et al. (2006); Motta et al. (2012); Fian et al. (2018); Popović et al. (2020); Fian et al. (2021)), as well as in the continuum emission of the accretion disk (Fohlmeister et al. (2008); Chen et al. (2012); Fian et al. (2016)).

From the microlensing variability in each different quasar image we can infer the accretion disk size of the quasar and the stellar mass fraction in the galaxy cluster where the images are located. In order to do so, we use the light curves from Muñoz et al. (2022). These light curves were acquired at the Fred Lawrence Whipple Observatory (FLWO) 1.2 m during 14.5 years. Given the length of the light curves, the time delay of image D was determined for the first time and it is the longest delay ever measured for a lensed quasar. Also, the time delays between images A, B and C were constrained more precisely than in the previous work of Fohlmeister et al. (2008). These three independent time delays are crucial to properly extract the microlensing variability in the light curves.

The newly measured time delay can also be used to refine the galaxy cluster mass model as in Forés-Toribio et al. (2022). The mass model was built incorporating these new measurements of time delays along with previous observations reported in Oguri (2010) (the images positions and flux ratios, the positions, luminosities and ellipticities of the galaxy cluster members, and the velocity dispersion of the brightest cluster galaxy). This mass model was able to precisely constrain the inner slope of the generalised Navarro-Frenk-White profile that was used to model the Dark Matter Halo of the cluster. The inclusion of the time delay of image D was directly related to this determination because its value strongly depends on the inner slope of this component (Kawano & Oguri (2006); Oguri (2010)). Apart from this determination, the convergence and shear at the quasar positions were computed from this mass model and are used to model the microlensing variability.

2. Methods and results

In order to estimate the microlensing variability, we need to remove the intrinsic variability produced by the quasar itself and the magnification that the lensing cluster provides to each image. To do so, we first smooth the light curves with a window of ten days to reduce the noise and we subtract one light curve to another after shifting them by their corresponding time delays, with this procedure the intrinsic variability is removed. The light curves that are going to be subtracted are fit with 5th order splines to interpolate their magnitudes when differentiated. To correct for the macro-magnification, we also subtract their magnitudes in infrared from Ross et al. (2009) which are expected to be only affected by macro-magnification. In this way, we compute the six independent microlensing differences between the images, i.e., A−B, C−B, D−B, A−C, D−C and A−D which are depicted in Figure 1.

With these residuals we construct histograms of microlensing differences by Monte Carlo sampling to account for observational errors. The histograms from observational data are compared with model histograms which depend on the half-light radius of the source, $R_{1/2}$, and the stellar fractions at each quasar image position, $\alpha = \kappa_*/\kappa$. These histograms are obtained from magnification maps computed with the Fast Multipole Method - Inverse Polygon Mapping (FMM-IPM) developed by Jiménez-Vicente & Mediavilla (2022). The magnification maps dimensions are $60{\times}60\ R_E$ with the convergence and shear from the mass model of Forés-Toribio et al. (2022) and with variable fraction of mass in stars. These stars have a mass of $0.3 M_\odot$ and are distributed randomly. Lastly the maps are convolved with different source sizes modelled as Gaussians.

The probability that a given set of parameters reproduce the observed microlensing differences, H, is:

$$P(H|R_{1/2}, \alpha_A, \alpha_B, \alpha_C, \alpha_D) \propto e^{-\chi^2/2} \tag{1}$$

where

$$\chi^2 = \sum_\mu \sum_i \left(\frac{h_\mu(i) - \tilde{h}_\mu(i; R_{1/2}, \alpha_X, \alpha_R)}{\epsilon_\mu(i)} \right)^2 . \tag{2}$$

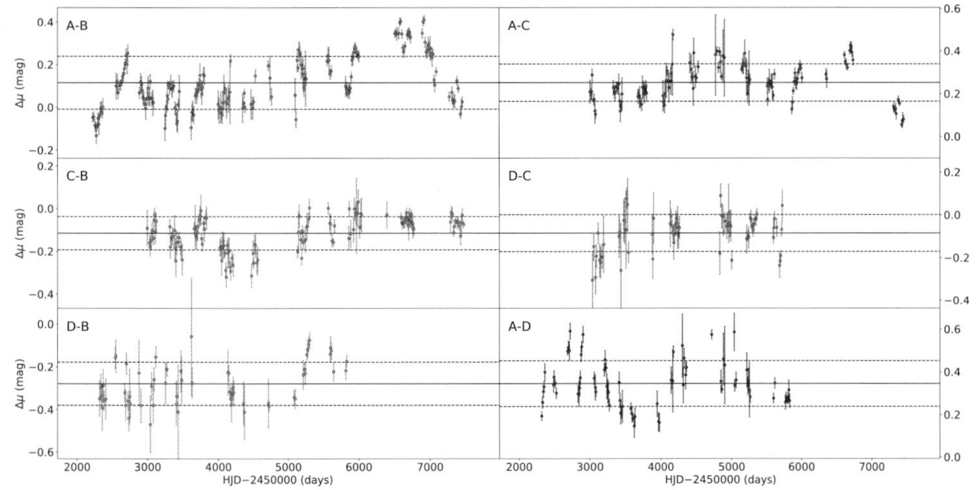

Figure 1. Microlensing differences between the four brightest quasar images. The solid black lines mark the average micro-magnification and the dashed lines show the standard deviation.

The first summation runs over the six microlensing differences and the second over the histogram bins. $h_\mu(i)$ is the ith bin of the normalised observed histogram, $\epsilon_\mu(i)$ is the error associated to that bin and $\tilde{h}_\mu(i; R_{1/2}, \alpha_X, \alpha_R)$ is ith bin of the difference of two model histograms for the given set of parameters (the half light radius of the source, $R_{1/2}$, the stellar fraction of the first image histogram, α_X, and the stellar fraction of the other image histogram, α_R).

We assign a logarithmic prior to the accretion disk half light radius and a uniform prior to the stellar fractions when their posterior probabilities are computed. We vary all the parameters but we keep the stellar fractions within a reasonable range according to the mass model of Forés-Toribio et al. (2022). With this procedure we can constrain the quasar accretion disk size to $R_{1/2} = 5.3^{+1.3}_{-0.7}\sqrt{M/0.3M_\odot}$ light-days at 2407Å in the restframe. However, the stellar fractions are mainly unconstrained as shown on the left panel of Figure 2.

Given that the source is properly constrained, we fix it to the central value and rerun the inference to attempt to determine confidence intervals for the stellar fractions. Under the hypothesis that the source size is $R_{1/2} = 5.3$ light-days, the probability distributions of the stellar fractions are presented in the right panel of Figure 2 and the central values and the 68% confidence intervals are $\alpha_A = 0.058^{+0.024}_{-0.032}$, $\alpha_B = 0.048^{+0.032}_{-0.014}$, $\alpha_C = 0.018^{+0.015}_{-0.018}$ and $\alpha_D = 0.008^{+0.033}_{-0.008}$.

3. Discussion and conclusions

In Figure 3 we compare previous size determinations at the same restframe wavelength and mean stellar mass with our inferred value. Our estimate is in tension with the determinations of Hutsemékers et al. (2023), Mosquera & Kochanek (2011) and Fohlmeister et al. (2008) who derived a smaller disk size. On the other hand, our value is compatible with the determinations of Fian et al. (2016), Jiménez-Vicente et al. (2014) and Motta et al. (2012). Given the length of the light curves, we achieve a tighter constraint on the source size than the majority of previous works.

Regarding the stellar fraction estimates, we can compare them with the stellar contributions from the brightest cluster galaxy (BCG) and the intracluster light (ICL) at the specific quasar positions derived from Kravtsov et al. (2018), DeMaio et al. (2018) and Henden et al. (2020) (see Table 1). According to these estimations, α_C and α_D lie

Table 1. Estimates of the stellar mass fraction from the brightest cluster galaxy and the intracluster light at the quasar image positions based on previous works.

BCG+ICL	K18	D18	H20
α_A	0.012 ± 0.007	0.006 ± 0.002	$0.012^{+0.022}_{-0.008}$
α_B	0.013 ± 0.008	0.007 ± 0.003	$0.013^{+0.025}_{-0.009}$
α_C	0.014 ± 0.009	0.008 ± 0.003	$0.011^{+0.022}_{-0.007}$
α_D	0.018 ± 0.010	0.038 ± 0.013	$0.022^{+0.028}_{-0.012}$

K18 = Kravtsov et al. (2018), D18 = DeMaio et al. (2018) and H20 = Henden et al. (2020).

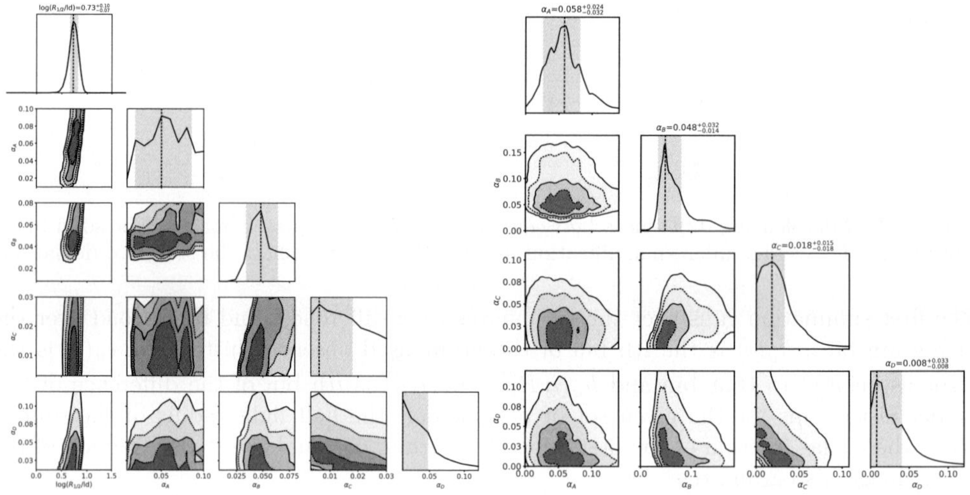

Figure 2. Probability distributions by pairs of parameters and marginalised distributions where the confidence intervals are reported at the 68% confidence level around the maximum. In the 2D plots, the 1-σ and 2-σ contours are marked with solid lines and the 0.5-σ and 1.5-σ are displayed as dashed lines. The left panel represents the joint inference for all five parameters ($R_{1/2}$, α_A, α_B, α_C and α_D) and the right panel shows only the inference for the stellar fractions when the source size is fixed to 5.3 light-days.

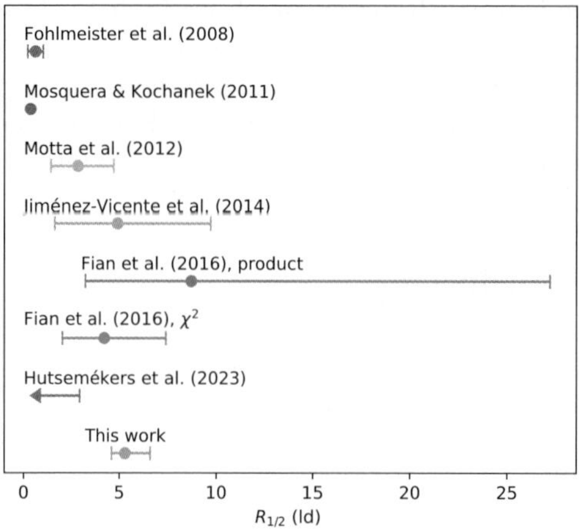

Figure 3. Quasar accretion disk size estimates for SDSS J1004+4112. The results are reported at $\lambda_{\rm rest}$=2407Å half-light radius in light-days for a mean stellar mass of $0.3M_\odot$.

around the expected values but at the image positions A and B we obtain slightly larger fractions. This may suggest the presence of undetected galaxy cluster members in those regions.

References

Chen, B., Dai, X., Kochanek, C. S., et al. 2012, *ApJ*, 755, 24.
DeMaio, T., Gonzalez, A. H., Zabludoff, A., et al. 2018, *MNRAS*, 474, 3009.
Fian, C., Guerras, E., Mediavilla, E., et al. 2018, *ApJ*, 859, 50.
Fian, C., Mediavilla, E., Hanslmeier, A., et al. 2016, *ApJ*, 830, 149.
Fian, C., Mediavilla, E., Motta, V., et al. 2021, *A&A*, 653, A109.
Fohlmeister, J., Kochanek, C. S., Falco, E. E., et al. 2008, *ApJ*, 676, 761.
Forés-Toribio, R., Muñoz, J. A., Kochanek, C. S., et al. 2022, *ApJ*, 937, 35.
Gómez-Álvarez, P., Mediavilla, E., Muñoz, J. A., et al. 2006, *ApJL*, 645, L5.
Henden, N. A., Puchwein, E., & Sijacki, D. 2020, *MNRAS*, 498, 2114.
Hutsemékers, D., Sluse, D., Savić, D., et al. 2023, *A&A*, 672, A45.
Inada, N., Oguri, M., Pindor, B., et al. 2003, *Nature*, 426, 810.
Jiménez-Vicente, J. & Mediavilla, E. 2022, *ApJ*, 941, 80.
Jiménez-Vicente, J., Mediavilla, E., Kochanek, C. S., et al. 2014, *ApJ*, 783, 47.
Kawano, Y. & Oguri, M. 2006, *PASJ*, 58, 271.
Kravtsov, A. V., Vikhlinin, A. A., & Meshcheryakov, A. V. 2018, *Astron. Lett.*, 44, 8.
Lamer, G., Schwope, A., Wisotzki, L., et al. 2006, *A&A*, 454, 493.
Mosquera, A. M. & Kochanek, C. S. 2011, *ApJ*, 738, 96.
Motta, V., Mediavilla, E., Falco, E., et al. 2012, *ApJ*, 755, 82.
Muñoz, J. A., Kochanek, C. S., Fohlmeister, J., et al. 2022, *ApJ*, 937, 34.
Oguri, M. 2010, *PASJ*, 62, 1017.
Popović, L. Č., Afanasiev, V. L., Moiseev, A., et al. 2020, *A&A*, 634, A27.
Richards, G. T., Keeton, C. R., Pindor, B., et al. 2004, *ApJ*, 610, 679.
Ross, N. R., Assef, R. J., Kochanek, C. S., et al. 2009, *ApJ*, 702, 472.

Strong Gravitational Lensing in the Era of Big Data
Proceedings IAU Symposium No. 381, 2024
H. Stacey, A. Sonnenfeld & C. Grillo, eds.
doi:10.1017/S1743921323003654

Probing the Structure of SDSS J1004+4112 through Microlensing Analysis of Spectroscopic Data

C. Fian[1], J. A. Muñoz[1,2], R. Forés-Toribio[1,2], E. Mediavilla[3,4], J. Jiménez-Vicente[5,6], D. Chelouche[7,8], S. Kaspi[9] and G. T. Richards[10]

[1]Departamento de Astronomía y Astrofísica, Universidad de Valencia, E-46100 Burjassot, Valencia, Spain. email: carina.fian@uv.es

[2]Observatorio Astronómico, Universidad de Valencia, E-46980 Paterna, Valencia, Spain

[3]Instituto de Astrofísica de Canarias, Vía Láctea S/N, La Laguna 38200, Tenerife, Spain

[4]Departamento de Astrofísica, Universidad de la Laguna, La Laguna 38200, Tenerife

[5]Departamento de Física Teórica y del Cosmos, Universidad de Granada, Campus de Fuentenueva, 18071 Granada, Spain

[6]Instituto Carlos I de Física Teórica y Computacional, Universidad de Granada, 18071 Granada, Spain

[7]Department of Physics, Faculty of Natural Sciences, University of Haifa, Haifa 3498838, Israel

[8]Haifa Research Center for Theoretical Physics and Astrophysics, University of Haifa, Haifa 3498838, Israel

[9]School of Physics and Astronomy and Wise Observatory, Raymond and Beverly Sackler Faculty of Exact Sciences, Tel-Aviv University, Tel-Aviv 6997801, Israel

[10]Department of Physics, Drexel University, 32 S. 32nd Street, Philadelphia, PA 19104, USA

Abstract. We aim to uncover the structure of the continuum and broad emission line (BEL) emitting regions in the gravitationally lensed quasar SDSS J1004+4112 through unique microlensing signatures. Analyzing 20 spectroscopic observations from 2003 to 2018, we study the striking deformations of various BEL profiles and determine the sizes of their respective emitting regions. We use the emission line cores as a baseline for no microlensing and then apply Bayesian methods to derive the sizes of the Lyα, Si IV, C IV, C III], and Mg II emitting regions, as well as of the underlying continuum-emitting sources. We find that the sizes of the emitting regions for the BELs are a few light-days across, notably smaller than in typical lensed quasars. The asymmetric distortions observed in the BELs suggest that the broad-line region lacks spherical symmetry and is likely confined to a plane. The inferred continuum emitting region sizes are larger than predictions based on standard thin-disk theory by a factor of ~ 4. We find that the size-wavelength relation is in agreement with that of a geometrically thin and optically thick accretion disk.

Keywords. gravitational lensing: strong – gravitational lensing: micro – quasars: general – quasars: emission lines – quasars: individual (SDSS J1004+4112)

---

1. Introduction

SDSS J1004+4112 is the first quasar known to be lensed by a foreground cluster of galaxies and was discovered in the Sloan Digital Sky Survey (Inada et al. 2003). The lensed quasar comprises four bright images at a source redshift of $z_s = 1.73$. These images are produced by a massive cluster at a lens redshift of $z_l = 0.68$, and the maximum separation angle is $14''.6$ between components B and C. The system has been monitored photometrically (Fohlmeister et al. 2007, 2008; Fian et al. 2016; Muñoz et al. 2022), and the time-delays for three of the quasar images (A, B, and C) were measured by Fohlmeister et al. (2007, 2008). A recent study by Muñoz et al. (2022) found a time delay of 2458.47 ± 1.02 days between the trailing image D and the leading image C, which is the longest delay ever measured for a gravitationally lensed quasar.

Optical microlensing, induced by stars in the cluster halo or nearby satellites, is a recognized phenomenon in the lens system SDSS J1004+4112. This effect has been used to estimate the size of the continuum-emitting source in the lensed quasar (Fohlmeister et al. 2008; Fian et al. 2016). Alongside the differential brightness variation of the images in photometric passbands, microlensing effects have also been identified in the spectral lines (Richards et al. 2004; Lamer et al. 2006; Gómez-Álvarez et al. 2006; Motta et al. 2012; Popović et al. 2020; Fian et al. 2018, 2021). Spectroscopic analyses of the individual components have unveiled significant differences in the emission line profiles. Particularly, component A displays an enhanced blue wing in various high-ionization lines relative to the other components, which has been interpreted as an indication of microlensing in image A. Recent research by Hutsemékers et al. (2023) has demonstrated that the distortion of the C IV emission line profile in SDSS J1004+4112 can be explained using simple broad-line region (BLR) models. Our study expands upon this by extensively analyzing existing spectroscopic data, including additional observational epochs and a wider range of spectral lines compared to Hutsemékers et al. (2023). Moreover, we undertake a comprehensive global analysis encompassing line cores, line wings, and the adjacent continuum, deepening our insight into the diverse light-emitting regions within the quasar.

2. Observations and Data Analysis

We have assembled a set of rest-frame UV spectra for the images A, B, C, and D of the gravitationally lensed quasar SDSS J1004+4112. Our dataset consists of 20 published spectra, spanning a 15-year period from February 2003 to February 2018. These spectra cover a range of typical high- and low-ionization lines found in quasars, including Lyα, Si IV, C IV, C III], and Mg II. In Figure 1, we show the average of these line profiles for each lensed image observed in at least two different epochs. We also include one, two, and three-sigma intervals to highlight the variability in the line wings and other emission species. We observe significant variability in the blue wings of Si IV, C IV, and C III] in image A, as well as in the red wing of Lyα and the shelf-like feature at $\sim \lambda 1610$ (blueward of He II).

Emission line cores arise from large, spatially extended regions that are not significantly affected by microlensing and intrinsic variability (Guerras et al. 2013; Fian et al. 2018). Measuring core flux ratios between lensed images provides a robust method for establishing a microlensing-free reference point. To remove the continuum from each image and emission line, we apply a linear fit to both sides of the emission line's continuum and subtract it from the spectrum. The core fluxes are then defined by a narrow interval centered on the peak of the line. Our inferred core flux ratios align well with infrared flux ratios reported by Ross et al. (2009) and show consistency with radio flux ratios observed by Jackson (2020), Hartley et al. (2021), and McKean et al.

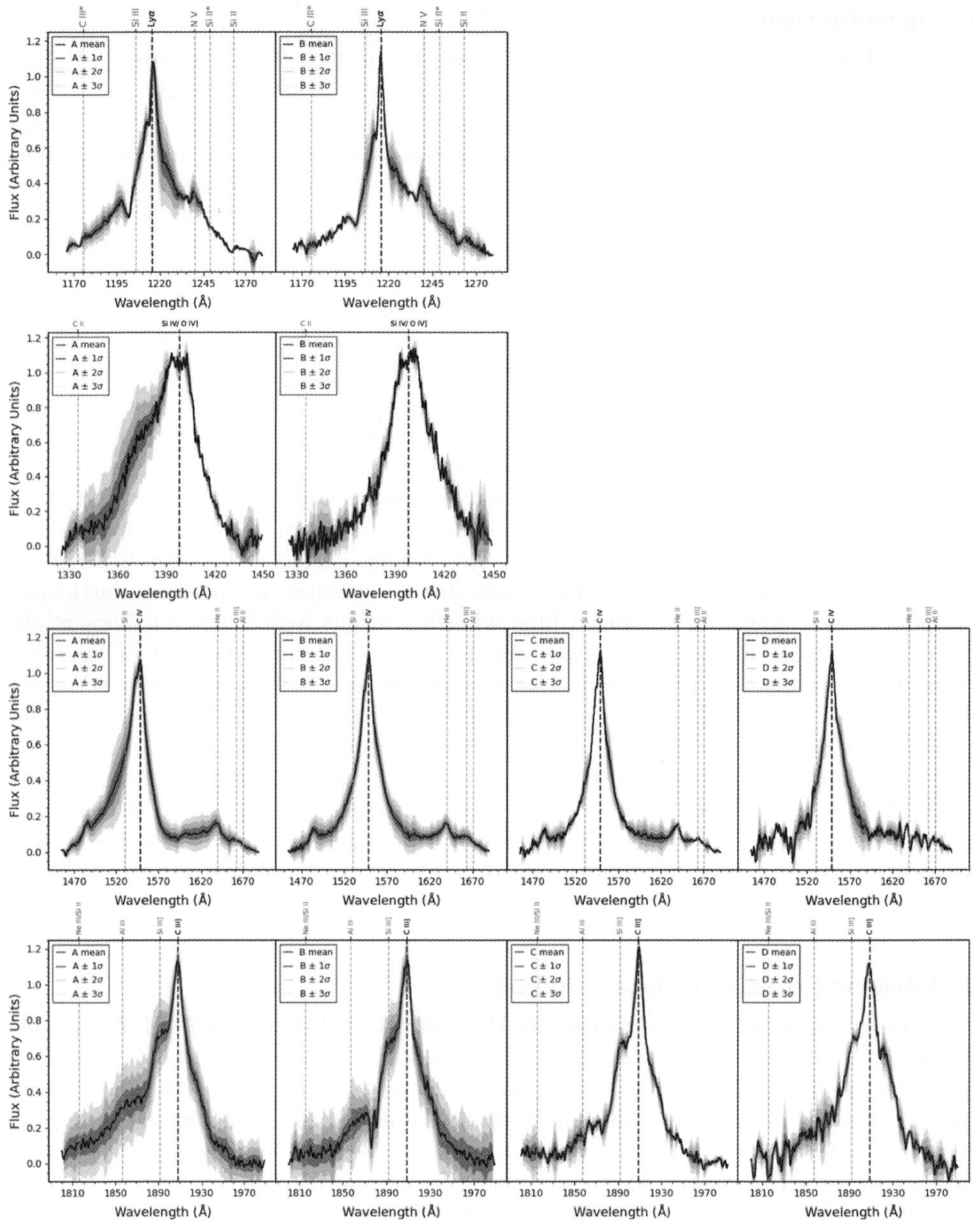

Figure 1. Average (black) spectra of the Lyα, Si IV, C IV, and C III] emission lines (from top to bottom). The one, two, and three sigma intervals for the images A, B, C, and D (left to right) are depicted in dark blue, blue, and light blue, respectively.

(2021). Microlensing, which is sensitive to the source size (where smaller regions experience higher magnifications), offers valuable insights into the structure and kinematics of quasar accretion disks. Quantifying the microlensing impact on the continuum requires disentangling it from macro-lensing magnification effects (stemming from the smooth lensing potential) and extinction. We correct for these effects by assessing the differences between the continuum adjacent to emission lines, $(m_x - m_B)_{cont}$, and the magnitude differences of emission line cores, $(m_x - m_B)_{core}$, between the images x (where $x = A, C, D$) relative to image B, $\Delta\mu_{cont} = (m_x - m_B)_{cont} - (m_x - m_B)_{core}$. Image B

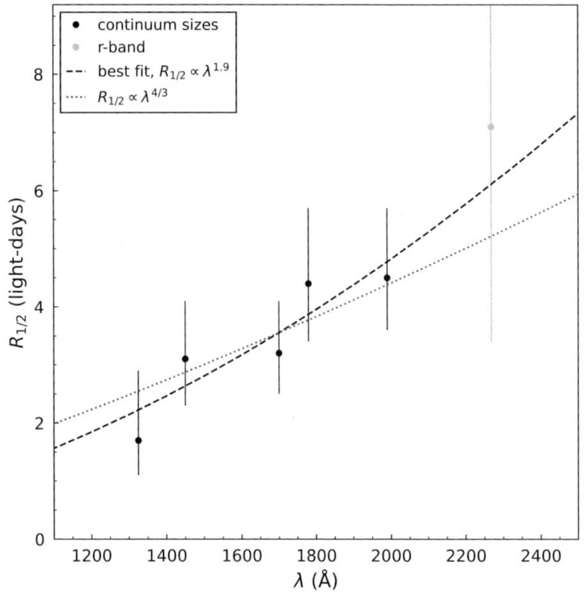

Figure 2. Continuum-emitting region sizes derived from microlensing, plotted against wavelength. The dashed black line shows the best fit to the data ($\beta \sim 1.9$). The red dotted line represents a fit with a fixed theoretical power-law index of $\beta = 4/3$.

is chosen as the reference as it is less affected by microlensing variability compared to image A (Hutsemékers et al. 2023). Additionally, it benefits from a more extensive set of observational epochs compared to images C and D. To determine the sizes of BEL-emitting regions, we normalize the continuum-subtracted spectra for all images and epochs to match the emission line core. We then assess the microlensing in the line wings ($\Delta\mu_{wing} = (m_x - m_B)_{wing} - (m_x - m_B)_{core}$) on either side of the emission line peak, corresponding to a velocity range of $3000 - 10000$ km s^{-1}.

3. Methods and Results

Given the observed differential microlensing in the wings and adjacent continua of various emission lines between lensed images, we can deduce the size of their emission regions. We employed a statistical approach, treating each microlensing measurement as a single epoch. All available epochs were used to calculate the joint microlensing probability to derive an average size estimate, as detailed in Guerras et al. (2013) and Fian et al. (2018, 2021). Inferring differential microlensing from the spectroscopic data of lensed quasars is complex due to time-delayed intrinsic flux variations between images. Deformations in the BELs caused by intrinsic variability can resemble microlensing, leading to potential inaccuracies in source size measurements. For the lensed quasar SDSS J1004+4112, this issue is particularly pronounced for images C and D, which have long time delays (~ 6.7 years) relative to other images in the system. To ensure accurate interpretation, our focus will be on the magnitude differences between images A and B, given their short time lag (~ 44 days), as these differences can most likely be attributed to microlensing.

Figure 2 illustrates the microlensing-based continuum sizes as a function of wavelength. To fit our size spectrum, we utilized the power-law index β — representing the disk's temperature profile — as an adjustable parameter, achieving the most optimal fit at $\beta \sim 1.9$. The inferred sizes and the physical model are consistent with the slope expected for an optically thick and geometrically thin accretion disk model ($\beta = 4/3$). While numerous microlensing studies have found sizes significantly larger than predicted

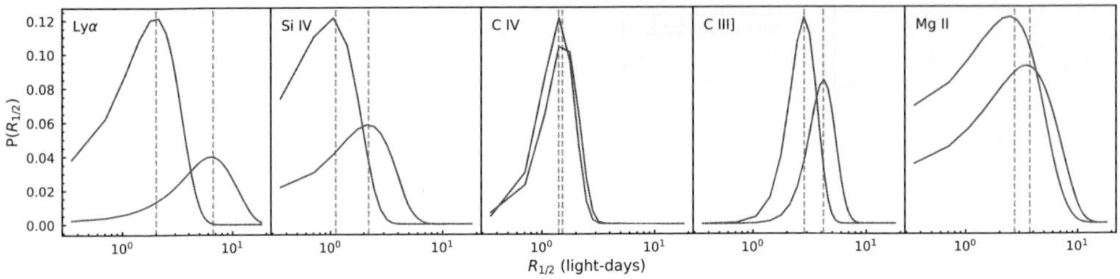

Figure 3. Probability density functions of the half-light radius $R_{1/2}$ for the blue and red wings of Lyα, Si IV, C IV, C III], and Mg II (left to right). The vertical dashed lines denote the expected emission region size.

by existing accretion disk theories, corroborating our results (~ 4 larger than theoretical predictions), they often report flatter size-wavelength relationships (Blackburne et al. 2011; Jiménez-Vicente et al. 2014). To affirm the accuracy of our results, we compared the r-band continuum size derived from our spectroscopic data with that acquired from 14.5 years of photometric monitoring data (Muñoz et al. 2022). The outcomes from the two distinct datasets align well, with $R_{1/2} = 7.1^{+7.4}_{-3.7}$ lt-days for the spectroscopic data, and $R_{1/2} = 5.3^{+1.3}_{-0.7}$ lt-days for the photometric observations (Forés-Toribio et al. submitted). In Figure 3, we display the probability density functions for the regions emitting the broad wings of Lyα, Si IV, C IV, C III], and Mg II. Intriguingly, we find that the sizes of the regions emitting the BELs are often very small, occasionally even smaller than the sizes associated with the optical continuum-emitting region. These results differ significantly (by an order of magnitude) from the BLR size estimates reported in Guerras et al. (2013), and they also contrast with the average BLR sizes determined for a sample of lensed quasars in prior studies (Fian et al. 2018, 2021). However, our inferred half-light radius for the C IV BLR is in agreement with the recent measurement by Hutsemékers et al. (2023) ($R_{1/2} = 2.8^{+2.0}_{-1.7}$ lt-days), lending credence to the reliability of our findings.

4. Conclusions

SDSS J1004+4112 stands out as an extensively studied lensed quasar, with abundant photometric monitoring data and spectroscopic observations available. Although recognized early as a lensed quasar with BEL deformation, understanding the line distortions and inter-component differences remained incomplete. With 20 spectroscopic observations, we thoroughly analyzed the Lyα, Si IV, C IV, C III], and Mg II lines, plus adjacent continua, and we constrained their emitting region sizes. We unveiled agreement in the line core ratios between the lensed quasar images, serving as a microlensing-free reference, and mid IR/radio ratios reported in the literature (Ross et al. 2009; Jackson 2020; Hartley et al. 2021; McKean et al. 2021). The inferred sizes of the continuum-emitting regions increase with wavelength, supporting the idea of disk reprocessing, and are larger by a factor of ~ 4 than predicted by a standard thin disk. This discrepancy is consistent with recent studies (Jiménez-Vicente et al. 2014; Fian et al. 2023), possibly resulting from a significant contribution of diffuse BLR emission to the observed continuum signals (see, e.g., Chelouche et al. 2019; Korista & Goad 2019; Netzer 2022; Fian et al. 2023). Notably, our spectroscopic r-band continuum size ($R_{1/2} = 7.1^{+7.4}_{-3.7}$ lt-days) matches well with the photometric data-based size ($5.3^{+1.3}_{-0.7}$ lt-days; see also Forés-Toribio et al. submitted). Using Bayesian analysis, we determined the emission region sizes for Lyα, Si IV, C IV, C III], and Mg II. Our findings show that these regions are a few light-days across, significantly smaller than average quasar BLRs (usually tens of light-days; Guerras et al. 2013; Fian et al. 2018, 2021). Intriguingly, some BEL-emitting region sizes are smaller than the optical continuum. A plausible interpretation for this surprising outcome could be that the BEL-emitting regions are located close to a caustic, while the accretion disk is farther away from it. Should the BLR, or a part thereof, be situated on or near a caustic while the accretion disk is located farther from it, the BLR would experience substantial magnification, whereas the accretion disk would undergo less magnification. While this situation is expected to be uncommon, it could conceivably account for the microlensing anomalies observed in this lensed system.

References

Blackburne, J. A., Pooley, D., Rappaport, S., et al. 2011, ApJ, 729, 34.
Chelouche, D., Pozo Nuñez, F., & Kaspi, S. 2019, Nature Astronomy, 3, 251.
Fian, C., Mediavilla, E., Hanslmeier, A., et al. 2016, ApJ, 830, 149.
Fian, C., Guerras, E., Mediavilla, E., et al. 2018, ApJ, 859, 50.
Fian, C., Mediavilla, E., Motta, V., et al. 2021, A&A, 653, A109.
Fian, C., Chelouche, D., & Kaspi, S. 2023, arXiv:2307.14824.
Fohlmeister, J., Kochanek, C. S., Falco, E. E., et al. 2007, ApJ, 662, 62.
Fohlmeister, J., Kochanek, C. S., Falco, E. E., et al. 2008, ApJ, 676, 761.
Gómez-Álvarez, P., Mediavilla, E., Muñoz, J. A., et al. 2006, ApJ, 645, L5.
Guerras, E., Mediavilla, E., Jimenez-Vicente, J., et al. 2013, ApJ, 764, 160.
Hartley, P., Jackson, N., Badole, S., et al. 2021, MNRAS, 508, 4625.
Hutsemékers, D., Sluse, D., Savić, D., et al. 2023, A&A, 672, A45.
Inada, N., Oguri, M., Pindor, B., et al. 2003, Nature, 426, 810.
Jackson, N. 2020, ApJ, 900, L15.
Jiménez-Vicente, J., Mediavilla, E., Kochanek, C. S., et al. 2014, ApJ, 783, 47.
Korista, K. T. & Goad, M. R. 2019, MNRAS, 489, 5284.
Lamer, G., Schwope, A., Wisotzki, L., et al. 2006, A&A, 454, 493.
McKean, J. P., Luichies, R., Drabent, A., et al. 2021, MNRAS, 505, L36.
Motta, V., Mediavilla, E., Falco, E., et al. 2012, ApJ, 755, 82.
Muñoz, J. A., Kochanek, C. S., Fohlmeister, J., et al. 2022, ApJ, 937, 34.
Netzer, H. 2022, MNRAS, 509, 2637.
Popović, L. Č., Afanasiev, V. L., Moiseev, A., et al. 2020, A&A, 634, A27.
Richards, G. T., Keeton, C. R., Pindor, B., et al. 2004, ApJ, 610, 679.
Ross, N. R., Assef, R. J., Kochanek, C. S., et al. 2009, ApJ, 702, 472.

Author index